Smart City 3.0

A new way of governing

By

Lorenzo Madrid & Linda Lee Bower

Table of Contents

1 INTRODUCTION

More than half of the world's population currently lives in urban areas. Further, urbanization is continuing. By 2030, about 200 cities[1] around the world will join the list of cities that have a metropolitan area population of over 1 million inhabitants.[2]

These concentrations of people have many benefits, but they also generate technical, physical, material, and social problems such as traffic congestion, waste management, safety, and security, among others. Urban governments seek to make their operations more efficient, save money, improve living conditions, and enhance the quality of life. Any help they can get from technology is welcome. Information and communications technologies (ICT) offer the possibility to achieve these goals. Investment in Smart City infrastructure is expected to improve the management of city operations and services.

Various forecasts for the Smart Cities market are vastly different, depending on what we count and the methodology the researchers use. Just a few estimates are:

- The Smart 2020 report expects that technologies and industries related to Smart Cities will rise to US$2.1 trillion by 2020.[3]

- A Market Watch study expects the global Smart Cities market to reach US$1.9 trillion by 2022.[4]

- A study put out by Markets and Markets predicts that the Smart Cities market will be worth over $717 billion by 2023.[5]

- Global Industry Analysts, Inc., forecasts that the global market for Smart City technologies will reach US$1.34 trillion by 2024.[6]

- The Smart Cities Association estimates that the global Smart Cities market will reach $3.5 trillion by 2026.[7]

Although these forecasts vary widely, they all have one thing in common: They all expect very rapid growth. Major market segments driving the growth include Smart Meters, Smart Lighting, Smart Traffic Control, and Smart Parking, among others.

This book presents an overview of the different aspects that need to be considered for the successful development of Smart Cities and also provides several examples of successes and failures. The aim is to create awareness of the difficulties one will face in implementing Smart City initiatives.

We will also present some recommendations that may help government agencies and practitioners to overcome the hurdles that exist on the path to achieving the ultimate goal of creating a smart, livable, and pleasant city.

We based this book on a great deal of research and the authors' extensive experience in implementing e-Government and Smart City projects around the world under diverse socio-economic conditions. It also draws on papers that the authors have presented jointly and individually at various conferences such as ICEGOV conferences in China, Egypt, Estonia, the United States, and Colombia, as well as the Samos Summit, IEEE-ISTAS, and other Smart City events in Croatia, Slovenia, the U.S., South Africa, and Brazil. We provide sources and references throughout.

Chapter 2 seeks to define what constitutes a Smart City and how it helps to achieve the objectives of municipal governments. It also traces the development of e-Government leading to Smart Cities, with the stages of evolution.

Chapter 3 looks at why projects fail, with lessons learned to help governments avoid the same mistakes. It points out the constraints that governments face and what is needed to succeed.

Chapter 4 looks at the requirements for a Smart City implementation, addressing the importance of having a strategic plan to provide a roadmap for the implementation of a Smart City. It discusses the vital elements in formulating a strategy, including policy, governance, resources, and a framework to organize the different components and ensure that everything works according to the plan.

Chapter 5 looks at the technology platform for building a Smart City, beginning with the architecture, which defines the organization of a system and identifying all the components, including the Internet of Things, the telecommunications infrastructure, the data processing infrastructure, the security layer, databases, middleware, and application areas.

Chapter 6 looks at the development impacts of Smart City programs. The discussion examines the benefits for several sectors, such as human capacity building, energy, health services, safety and security, and aspects such as inclusivity and sustainability, among others.

Chapter 7 examines the evaluation of Smart City projects, with both economic metrics and non-economic metrics.

Chapter 8 considers the different business models most commonly used to implement Smart City projects.

Chapter 9 presents case studies of various cities that are well advanced in developing a Smart City.

The Annex provides a taxonomy that, based on a literature search, reviews, and categorizes Smart City initiatives in ten areas.

2   WHAT IS A SMART CITY?

To define a Smart City is not an easy task. There are many definitions available, and even after many years of the first mention of Smart City in the literature, we still lack a formal definition. The "Smart City" is a complex and heterogeneous concept that involves the use of ICT-based innovation in the urban space to improve the quality of life. Attaining the Smart City is a process rather than a specific technical solution, and it is indeed "A New Way of Governing."

Being such a fluid concept, the definition of a Smart City has been the subject of many academic works. In a recent paper presented at the ICEGOV/2019 in Melbourne, Australia, "A Taxonomy of Smart Cities Initiatives," the authors collected thirteen different definitions in the related literature.[8]

Using a simple statistical analysis of the wording from such descriptions, we can pinpoint the main objectives for a Smart City. The resulting word cloud picture from such analysis is in Figure 1

As we can see, the topics related to technology, as well as the ones associated with the quality of life for the citizens and the urban environment, stand out from the image.

Figure 1: Word Cloud representing Definitions of Smart City

Source: Authors

Based on these results and understanding that a Smart City is a complex entity, for this book, we will consider (see Figure 2) a Smart City as one that uses ICT-based innovation in the urban space seeking to:

Figure 2: What a Smart City Seeks to Achieve

Increase	the public administration capacity in the provision of public services[*]
Improve	citizens' quality of life by offering better living conditions
Promote	democracy by involving citizens in the government
Promote	sustainable economic development and city attractiveness (Dameri and Benevolo, 2016)

(*) This topic is where eGov and Smart City share common objectives. When we talk about providing better city services, they include the traditional eGov solutions, such as the one related to tax and fee payments, record keeping, requests for city services, among many others.

Many of the issues related to Smart City development are very similar to the ones from an e-Government implementation. In an early stage, they both developed similar systems in parallel—primarily online information and services. For example, at the national level, people can apply for social benefits online and file their tax returns online. At the local level, people can apply for a building permit or a business license, among other online activities.

Historically, as governments began offering e-services to their citizens, we found that technology could also be used in a broader sense to provide not only better citizen services, but also to improve quality of life for the city's inhabitants and promote social and economic development. Rather than limiting these initiatives to automating government workflows and reducing bureaucratic red tape, academics, engineers, policymakers and other stakeholders saw the opportunity to create a new model to operate and manage a city, based on the promises of the latest IT technology such as IoT (Internet of Things), advanced telecom networks, and the cloud.

The Smart City diverges from e-Government in significant part due to the needs generated by the daily activities of populations concentrated in urban areas—getting to work and the need for water and utilities, among others. These activities create the need for public street lighting, traffic control, quality control of the public water supply, and waste management, among others.

Smart Cities and e-Government are an intertwined concept. They share many goals, objectives, and technologies while at the same time, they may have their exclusive realms. A common goal is to reduce costs and consumption of resources while enabling the government to engage more effectively and actively with its citizens. On the other hand, to intelligently automate some city functions may not be part of government duties, but it makes the city smarter.

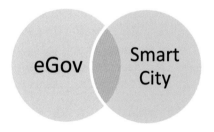

Since they are similar endeavors, Smart Cities and e-Government suffer the same kind of implementation challenges. The evolution of the Smart City into different maturity levels is comparable to that of e-Government. Therefore, lessons learned from e-Government deployments should apply to Smart City deployments. Thus, in this book, we will recall lessons from eGov implementations, to create awareness for what to expect on Smart City deployments.

Also, as it happens with e-Government, *a Smart City is not just a target or an artifact to be built. It is instead a complex process that requires long term vision, commitment, and political will from city leaders and other stakeholders.*

However, probably the essential fact about a Smart City implementation is that the process *requires a new way of governing*, as we will discuss throughout this book.

2.1 Objectives of Municipal Government

It takes many complex systems to operate a modern city, such as water supply, roads, mass transit, education, and social services, among others. Municipal governance involves planning, execution, and control. The implementation of ICT solutions can enhance the efficiency of this process, improve transparency, and increase stakeholder participation.

The primary objectives of a municipal government are to operate city functions efficiently and serve the citizens. Smart City solutions can enhance the performance of municipal operations, enable the citizens to participate more fully in the government, and enhance the quality of life.

Improving the quality of life involves making municipal systems more responsive to citizen needs, e.g., reducing traffic congestion, ensuring that the public water supply is safe, and having schools that produce literate and employable graduates. The better these systems are, the better the city is served.

By employing Smart City technologies, tools, and solutions, we can set realistic targets, and outcomes can be measured. Further, to best serve the citizenry, citizens themselves should be included in the decision-making process, along with other stakeholders. A Smart City platform--running under a cloud computing environment and capable of handling Big Data, Open Data, Legacy Data, managing multiple digital services, and autonomously making decisions based on pre-defined business rule--can progress toward the development and sustainability goals that a city sets.

2.2 Evolution of the Smart City Concept

The idea of applying ICT to improve municipal functions represents the initial step of a Smart City, but the concept has evolved into a broader set of requirements and definitions. Through the years, the idea of e-

Government has progressed through three stages, and the same thing is happening with the concept of a Smart City. The comparative evolution of evolution, for e-Government and Smart Cities, can be summarized as seen in Figure 3:

Figure 3: Comparative Evolution of e-Government

	Gov 1.0 e- Government	Gov 2.0 Platform Govt.	Gov 3.0 Smart Govt.
Concept	• A Government that works well	• Open government to people	• Evidence-based decision making
Goal	• Process Innovation	• Governance Innovation	• Policy Innovation
Resource	• IT Systems	• Web & Apps	• Data

The original concept was simply to use ICT solutions to improve the performance of government functions. This initial phase can be considered eGov 1.0. This stage has yielded significant results. For example, it enables a faster process for government paperwork, improves tax collection and reduces tax evasion, and expedites citizen services requests to the government.

As expressed by the United Nations:

> E-government can provide the necessary tools to enable policy integration not only across economic, social, and environmental dimensions but also among various sectors, subsectors, and programs. It can help "siloed" government institutions to join forces to pursue shared objectives through whole-of-government approaches. E-government can also offer opportunities to re-engineer existing decision-making processes and information flows.[9]

Then capabilities and expectations evolved, as expressed by Mauro Rio:

> Digital Government is a new way of organization and management of public affairs, introducing positive

transformational processes in management and the structure itself of the organization chart, adding value to the procedures and services provided, all through the introduction and continued appropriation of information and communication technologies as a facilitator of these transformations.[10]

From directly applying ICT solutions to make existing processes better, the methods and organizations are themselves transformed so that they can perform better, which means that they serve the citizen better. The citizenry is brought into the process as a participant to enable open and collaborative governance. That stage can be considered eGov 2.0. At both the national and local level, citizens can participate more fully in government. The availability of Open Data, enabling an Open Government process, is a vital constituent of this phase.

Now the evolution has continued to encompass policy; this brings us to eGov 3.0, which offers integrated solutions and new services. However, reaching an eGov 3.0 level is not an event by itself; it is a journey that takes time to achieve; along the way, we make mistakes and learn from them.

ICT solutions in government have enabled progress in virtually all sectors— not only public services, but education, healthcare, agriculture, and science, among others. Not only are services improved and enhanced, but more people have access to them.

The definition of e-government is generally associated with the use of ICT to make government functions more efficient and cost-effective. Governance, on the other hand, refers to the process by which society solves problems and meets the needs of the citizens. Ntulo et al. note that e-Governance refers to the whole spectrum of the relationship and networks within the government regarding the usage and application of ICT. It is a form of e-business that involves the processes and structures required to deliver electronic services to the citizenry, collaborate with partners, and conduct electronic transactions. E-Government can be a significant enabler of good governance practices. In addition to improving efficiency in the delivery of government services, e-Government can "simplify compliance with government regulations, strengthen citizen participation and trust in government…"[11]

2.3   Smart Cities Generations

eGov and Smart Cities can share many systems and objectives; for that reason, we believe it is essential to point out the parallels of both initiatives.

To illustrate this comparison, Figure 4 presents the concept for the different Smart City generations, their goals, and objectives, using the same framework used above for eGov.

Figure 4: Evolution of Smart City

	Smart City 1.0	Smart City 2.0	Smart City 3.0
Concept	• A City that works well	• An Inclusive & Responsive City	• A City that Thinks
Goal	• Operation Automation	• Services Integration	• Autonomous Management
Resource	• IoT Systems	• Web & Apps	• Data Analytics & Artificial Intelligence

The original intention of e-Government was to improve the efficiency of government operations. Smart Cities today have the same ambition. As the OECD points out,[12] Smart City solutions support and foster better government—better policy outcomes, higher quality services, more effective collaboration among agencies, and enhanced engagement with the citizen.

One of the aspects of service improvement is that citizens spend less time and resources obtaining a service. Rather than having to visit a government office in person and wait in line, services are now available online. By improving citizen services using eGov, we are also enhancing the quality of life and fostering economic development, two objectives of a Smart City.

E-government can help to achieve the intended outcomes of specific policies, contribute to economic objectives, and even contribute to reform. For example, sharing information in the health sector can improve the use of resources and patient care. More effective programs and improvements in business productivity have a positive impact on the economy. The technology fosters reform in many areas by improving transparency, facilitating information sharing, and revealing internal inconsistencies.

We can enhance engagement with citizens by promoting open and accountable government. These actions can also help to prevent corruption and foster trust.[13]

The conclusion is that the original objectives of eGov have become some of the building blocks of a Smart City!

2.4   Smart City Solutions Taxonomy

Usually, the first initiatives to deploy a Smart City come from an isolated need to address a specific city situation. Most of these initiatives revolve around installing Smart Street Lighting, using Street Sensors to control car parking spaces or waste collection systems. Nevertheless, there are many other possibilities for the deployment of these specific solutions, also called Vertical Solutions. These vertical solutions tend to address issues related to efficiency in the areas of energy, cost, or management. However, there are many more initiatives that can fit under the label of Smart City Solution.

In a recent study presented at the ICEGOV-2019 conference, "A Taxonomy of Smart Cities initiatives,"[14] the authors identified ten main categories for Smart City Initiatives and 85 specific actions or solutions that have been deployed or are in the process of deployment. Below, we show the diagram representing these solutions in Figure 5, including the main actors that can be involved in the process. The table with the 85 actions or solutions is available in the annex of this book.

Figure 5: Categories of Smart City Solutions

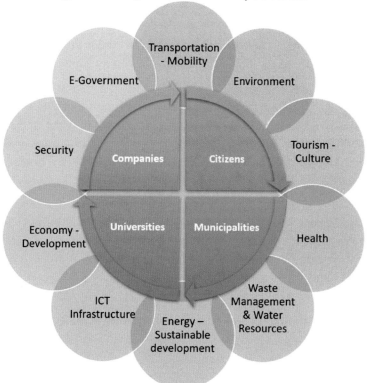

As the authors recognize, these categories and actions are not set in stone, but should be a dynamic classification; additional classifications and new activities will undoubtedly be necessary as the development of Smart Cities continues to occur. To that extent, the model is a living entity that needs to be continuously updated.

This living classification is available on an online mindmap,[15] where changes and updates will be posted, once accepted by the community of researchers.

2.5   Smart City Standards

The setting of technical standards is the cornerstone to enable the proper construction of an engineered solution. Of importance are IT standards **and** the tools to allow converting data formats from one standard to another.

Further, standards change and evolve, therefore limiting the adoption to a small set of criteria may prevent different components of a system—devices, applications, and networks—from communicating and working together to perform a process. Open standards are generally a good option, as they allow for interoperability, scalability, and market support, among other characteristics.

However, we shall not discard proprietary standards. The adoption of widely accepted market standards will facilitate implementation, adoption, and cross-platform support. Although open standards may initially look more attractive, many times their support by industry players may not be extensive, and frequently proper tools are lacking to accomplish the required tasks.

The standards should enable interoperability in the multiple aspects of a system: Technical, Organizational, Semantic, and Political. Even today, most governments that seek to implement an interoperability framework generally focus only on technical interoperability. It is reasonable to begin with technical interoperability because it is the easiest. However, the process should not stop there or be limited by it. These elements and standards for interoperability apply to e-Government, and Smart Cities follow. However, technical demands may differ substantially. Smart Cities will require new standards far beyond the ones needed for e-Government. For formulating a broader interoperability framework or taking it to a higher level, developing an Enterprise Architecture is an excellent alternative.

An early White Paper by the Danish Government on its enterprise architecture sets forth accepted principles for developing its interoperability: security, openness, flexibility, and scalability. The government should adopt a service-oriented architecture model, where the design for IT solutions is composed of service modules that have well-defined interfaces with each other and to legacy systems. The modules offer and use each other's services. *The service-oriented architecture does not itself establish any particular technical standards.*[16]

The adoption of standards had been a critical issue related to eGov interoperability. However, the focus for eGov standards is currently not as important as it was a few years ago. They were necessary for the initial phases of eGov, as they are necessary now for the initial stages of Smart Cities. Standards for Smart City products and solutions are still in their infancy. Nevertheless, experience has shown us that there is a maturity process for standards adoption, and usually they converge to a set of market accepted standards. If more than one standard becomes popular, middleware tools can take the burden from conversion needs.

Some countries have already initiated the development of Smart City standards such as Spain, Austria[17], and the UK.[18] In Spain, for example, AENOR is the Spanish standardization body responsible for national standards, and the organization represents Spain at the International and European Standardization Organizations. In 2019, AENOR established a Technical Committee on Smart Cities[19] (CTN178) as an initiative of the Secretary of State for Telecommunications and the Information Society in the Ministry of Industry, Energy, and Tourism. The purpose of this committee is to promote, streamline, and optimize the implementation of Smart Cities in Spain. It seeks to formulate a definition of a Smart City and to ensure interoperability of solutions. The standardization structure covers the elements seen in Figure 6.

Figure 6: Structure of Standardization Elements

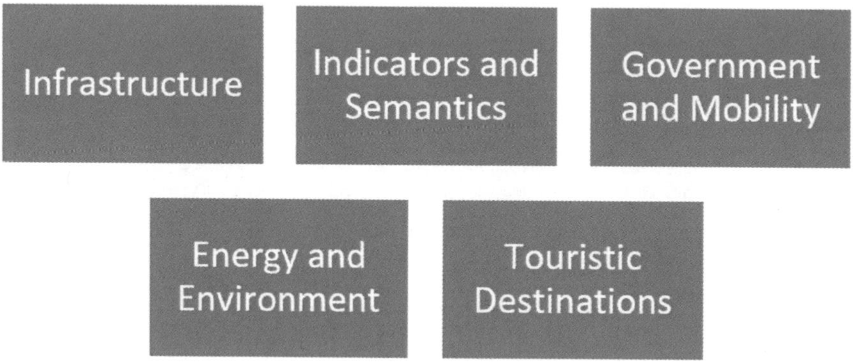

Infrastructure

Indicators and Semantics

Government and Mobility

Energy and Environment

Touristic Destinations

In the United States, the National Institute of Standards and Technology (NIST) is developing an IES-City Framework (Internet of Things Enabled Smart City Framework) and has several working documents[20] about these topics. We will come back to them in a future chapter.

2.6   Smart Territories

Some Smart City vendors are already promoting the idea of "Smart Territory," which aims to provide the same solutions across neighboring cities. The objective is to facilitate the life of citizens who need to cross the city-borders in their daily lives. The answer is to develop shared services and applications, and also to provide easy management and economies of scale for all cities involved.

Some metropolitan areas are already engaging in such deployments, which require an additional level of coordination and political will among the different stakeholders representing the cities or municipalities involved.

EXAMPLES OF SMART TERRITORIES: DALLAS, ISTANBUL, AND SYDNEY

Dallas

- The Dallas metropolitan area covers numerous municipalities such as Garland, Richardson, Plano, Addison, and Irving, among others. The DART system (Dallas Area Rapid Transit), which encompasses buses and light rail, serves these various municipalities; With a DART pass on a Smartphone, a passenger can travel on any bus or train throughout the system. It takes substantial cooperation and collaboration among the various authorities involved.

Istanbul

- The Istanbul Metropolitan Municipality includes various communities that surround the City of Istanbul, and they all involved in the Istanbul Regional Plan and participate in Istanbul's Smart City program.

Sydney

- The Sydney, Australia, metropolitan area, similarly extends to several surrounding cities, and they are participants in the Sydney Digital Strategy.

This idea can even extend to the national level. For example, as seen in the case study on Sydney, Australia (see Chapter 9), Sydney's digital plan is part of an overall development plan for the city, and it also fits into a national program to develop sustainable cities.

3 WHY DO PROJECTS FAIL?

Understanding why government projects have failed in the past can help to prevent making similar mistakes when we begin a Smart City initiative.

Many governments, municipalities, and national governments around the world seeking to improve their operations invest in technology. However, frequently, these investments do not yield the expected benefits. It is not merely a matter of installing equipment and systems. Obstacles to success can include outdated policies, constraints on resources, and the reluctance of bureaucrats to make changes. Understanding why such projects have failed can help us to prevent failures in Smart City projects. For that reason, we will review some failures in governmental projects and see what lessons we can learn to avoid a similar fate for Smart City initiatives. Whatever the level of government, the same principles are needed to achieve success. Lessons learned from a national-level project can apply to a local-level project.

As Pardo and Burke point out,[21] city governments need to deal with multiple and diverse stakeholders, high levels of interdependence, and competing values, as well as a complex social and political environment. To govern effectively in such a situation, government officials need information and knowledge. However, a typical hierarchical bureaucracy, with separate departments for different functions, is not conducive to information sharing or collaboration among departments.

3.1 The Constraints

Implementing a Smart City solution requires a great deal of interaction among existing ICT systems and new ones to be deployed. These interactions require advanced interoperability capabilities for the e-Government Systems, and we have learned that this process is complicated and cumbersome. Pardo and Burke summarize the constraints of deploying interoperable e-government services, among others:

- A lack of experience leading network forms of government
- Insufficient or no cross-boundary governance structure

- A lack of policies that allow new models for innovative resource allocation
- A lack of policies that engender investments in the principles of scalability and sustainability of solutions
- A focus on crisis-oriented response

What is needed is a network form of organization, where sharing of information, resources, and authority, as well as negotiation and work collaboration, cut across department and program lines, which enables government interoperability. Pardo and Burke define government interoperability as "the mix of policy, management, and technology capabilities needed by a network of organizations to deliver coordinated government programs and services." Government interoperability can enable the government to deliver coordinated government programs and services, as well as to share resources and knowledge more effectively. Achieving government interoperability has become an essential goal for many governments, and as such, it is a crucial element for successful deployment of ICT solution to address the needs of a Smart City.

They note that true government interoperability also requires organizational changes, new practices, and new cross-boundary relationships. However, this aspect is beyond the scope of this book. Although vendors and governments usually focus on the contribution of technology to achieving Smart Cities results, in this book, we point out other pieces of the puzzle.

ICT has already significantly improved the efficiency of certain governing functions. For example, many cities have an online procurement system. Vendors that do business with the city register once in a Procurement System, and they are then eligible to bid on any solicitation. Licitation documents are available on the web, bid submissions are online, and winners are announced and paid through the system. Citizens can also pay property taxes online or check out books in their local library using internet applications. However, each of these e-Government programs is still for the benefit of a single department within the city. True city e-Government interoperability implies the ability of multiple systems to exchange information with each other and use the previously collected information. The more that we can do this in the context of an IT system in a city, the smarter the city becomes.

In traditional government, the citizen visited a government office, filled out forms, and waited in line until his or her turn to get the attention from a clerk. When multiple services were involved, the citizen would have to fill out different forms for each agency, all of them containing some of the same information. This process continues to be the case in many cities today. In the next stage, certain individual agencies may streamline their processes to eliminate filling out forms manually and waiting in line; but other agencies still use the same inefficient practices. At the level of connected services, now all agencies may have adopted ICT solutions to serve their customers, but usually, each agency follows its practices, and the various agencies do not share information. Finally, we reach a seamless service level where all agencies are online, share information, and cooperate to serve the citizen.

Interoperability is the key to expanding the functionality of existing technology, and it is the foundation to help public sector workers to be effective and efficient in serving citizens.

Conceptually, we need to extend the concept of e-Government interoperability to other areas of a city. For instance, managing the city transit system or the traffic lights is also part of the city manager's job, and these functions need to be integrated and able to interoperate. So, the concept of interoperability goes beyond the needs of traditional e-Government. Smart devices, sensors, cameras, and other IoT appliances or CPS (Cyber-Physical Systems) will be part of the elements used to create a Smart City; and all the information coming from/to these devices needs to communicate with multiple systems, across different protocols and standards, and also serve multipurpose objectives across the participating agencies. Therefore, broad interoperability capabilities will be essential to achieve the desired result.

To fully harness the power of Smart City technologies, government agencies need to overcome the problem created by the existing ICT solutions that cannot communicate with each other. In many cases, we see a patchwork of ad-hoc solutions to mitigate their data communication issues. For that reason, interoperability also constitutes a key element to build and operate a Smart City so to achieve the goal of enhanced government/city efficiency and transparency in the delivery of public services to all citizens.

3.2   The Epitome of Failure

The two histories below illustrate the problems that can arise when policies are not apparent, different stakeholders do not communicate, and the organization does not provide sufficient supervision. One concerns a public transportation system in Santiago Chile. The other is regarding the development of a Website for a program at the federal level in the United States. Both situations bring us relevant lessons for the development of a transport service, a Website, or any other municipal service.

3.2.1   TranSantiago Public Transit System Collapse

Santiago, the capital of Chile, has a population of over 6 million. For many years, the elements of public transport in Santiago operated separately. The Metro subway, built in the early 1970s, received operating subsidies from the government and worked efficiently. The bus system was composed of hundreds of privately-owned bus lines. Free enterprise and competition created a network of bus routes that took people where they wanted to go and kept fares low. However, the competitive incentives gave rise to specific problems that the government wanted to remedy. The government imposed "reforms" intended to enhance the quality of public transport and reduce travel times for riders. What gave the reformed system the characteristic of a Smart City project is the electronic farecard now used throughout the system for both Metro and busses.

As Munger[22] tells the story, in 2007 the government decided to reform the public transit system, to be known as TranSantiago. Under this reform, the government consolidated the bus service into ten zones, each one under a concession. However, the changes suffered from poor planning and execution and caused even worse problems, including lower service quality, longer travel times, rider dissatisfaction, and financial losses.

Many problems arose. Among them, many routes changed so that rather than being able to take one bus from origin to destination, passengers had to take a bus to a Metro station and transfer to the Metro, and then transfer back to a bus for the remainder of the trip. Whereas before people were generally satisfied with service, now many people thought that the service was poor. Further, the Metro had generally been considered to be fast and dependable, but it became severely overcrowded.

> **In reality, the different system components such as infrastructure, support systems, financial administration and control measures, were ill-equipped to meet the challenges involved in a simultaneous launch.**
>
> - Global Mass Transit Report
> "TranSantiago reforms, Chile"
> December 1, 2017

The first ten years of TranSantiago represented an investment of about US$15 billion,[23] and it was one of the most ambitious transportation projects undertaken in a developing country.

A new procurement in 2017 brought new operating contracts, which contain incentives designed to foster excellent performance. The basis of the new system relied on three principles: better service quality, increased competition within the system, and much better continuity of services. Other improvements include superior bus technology, improved routing, better fleet management, improved infrastructure, and improved payment methods. For example, half of the bus fleet (some 3,300 busses) is being replaced, primarily with electric buses, at the cost of some US$500 million.

Development and improvements continue. For example, there is an app that enables passengers to know when their bus will arrive at the stop.

In 2019, the name was changed from TranSantiago to "**Red Metropolitana de Movilidad**" (Metropolitan mobility network). It appears that the main reason for the name change was to leave the poor reputation of TranSantiago behind.

3.2.2 HEALTHCARE.GOV FIASCO

Even in an advanced country, governments may have difficulties in achieving a successful Smart City or e-government project. For example, the United States government had lagged in this area and has missed opportunities due to poor management of technology investments. IT projects had frequently suffered cost overruns amounting to hundreds of millions of dollars and taking years longer than necessary to deploy, resulting in systems that were obsolete by the time they finished.[24] The U.S. Federal Government has frittered away hundreds of millions of dollars with poor outcomes.

The U.S. *HealthCare.gov* program epitomizes this situation, which demonstrates the results of a lack of mission definition, leadership, collaboration, proactivity, and accountability.

During the presidency of Barack Obama, the Affordable Care Act (ACA), also known as Obamacare, was passed into law. The system was intended to enable everyone to have healthcare insurance, and the Website healthcare.gov was created to support ACA. The purpose of this Website was to allow people to search for and find suitable insurance options to meet their needs at affordable prices.

Launched in 2013, the Website Healthcare.gov was supposed to be able to handle tens of thousands of visitors at once; but in a trial run just a few days before going online, the system could handle only a few hundred users. In spite of this warning, the launch went forward. On opening day, October 1, 2013, as a couple of thousand people visited the site and tried to go through the sign-up procedure, the site crashed. For some time after, users were frustrated by Website outages and technical malfunctions.

Cost overruns plagued the program, even after a change in the contractor. As documented by the Office of the Inspector General of the U.S. Department of Health & Human Services in its report,[25] the Obama administration spent some $840 million on the website www.HealthCare.gov. The GAO study covers five years from development, through the launch, corrective action, and implementation for the second enrollment period.

Characterized by technical complexity, the undertaking had a fixed deadline, but uncertainty about mission, scope, and funding. Many missteps throughout the process contributed to the disastrous launch. Above all, the Inspector General found numerous management failures, lack of adequate oversight, and contracting problems throughout the entire process. However, the most critical was the absence of clear leadership. In particular, as seen in Figure 7:

Figure 7: Failures of Leadership in Obamacare Program

 Too much time was spent on developing policy, leaving too little time to develop the Website.

 Poor technical decisions were made.

 There was a failure to adequately manage the key development contract.

The Centers for Medicare & Medicaid Services (CMS), the entity responsible for building the federally run insurance marketplaces, suffered structural, cultural, and tactical deficiencies. There was little coordination between policy and technical work, as well as a lack of communication. Leadership failed to heed warnings, apparently not wanting to hear bad news, and continued on the path that led to failure.

- When the launch of the Website revealed the significant problems, CMS and the contractors finally focused on corrective action. They reorganized the work and restructured the project so that CMS staff and contractor personnel collaborated. The HealthCare.gov Website recovered, and within two months it was available for high volume usage. Improvements continued, and as of February 1, 2016, over 9.6 million consumers had obtained coverage.[26]

The government learned some hard lessons from this appalling performance, summarized in Figure 8.

Figure 8: Lessons Learned from HealthCare.gov Disaster

①	**Leadership** Assign clear project leadership to provide cohesion across tasks and a comprehensive view of progress.	⑥	**Communication** Promote acceptance of bad news and encourage staff to identify and communicate problems.
②	**Alignment** Align project and organizational strategies with the resources and expertise available.	⑦	**Execution** Design clear strategies for disciplined execution, and continually measure progress.
③	**Culture** Identify and address factors of organizational culture that may affect project success.	⑧	**Oversight** Ensure effectiveness of IT contracts by promoting innovation, integration, and rigorous oversight.
④	**Simplification** Seek to simplify processes, particularly for projects with a high risk of failure.	⑨	**Planning** Develop contingency plans that are quickly actionable, such as redundant and scalable systems.
⑤	**Integration** Integrate policy and technological work to promote operational awareness.	⑩	**Learning** Promote continuous learning to allow for flexibility and changing course quickly when needed.

Source: U.S. Department of Health and Human Services, Office of Inspector General, *Healthare.gov*, Washington, D.C., February 2016, p. ii.3

The lessons of this failure apply to a Website developed for municipal service or any other level of government.

> **The failure to manage high risks leads to total failure in technology-driven public sector projects. 85% percent of IT projects fail because of the challenges by non-technical aspects of innovation in large part—policy, organization, and management related risks). Common reasons include poor planning, weak business case, lack of top management support, lack of leadership, lack of professional skills, misalignment between organizational goals and project objectives, vulnerability to policy swings, too much technology-driven enthusiasm, and political hyper-activism.**
>
> - Taewoo Nam and Theresa A. Pardo
> "Smart City as Urban Innovation: Focusing on Management, Policy, and Context,"
> Center for Technology in Government, University at Albany
> State University of New York
> September 2011

4 REQUIREMENTS FOR A SMART CITY IMPLEMENTATION

Implementing a Smart City is a Herculean task and prone to failure. As with any other complex process, a successful Smart City program requires, first of all, a vision for the desired outcomes and the political will to accomplish it. A Smart City implementation needs adequate public policies and a governance plan to execute the programs, along with a portfolio of projects to achieve the stated goals and objectives. Further, a Smart City calls for substantial resources, which we can aggregate in three main categories: technology, financial, and human. All of these elements must be wrapped up in a framework encompassing the organization, the information systems, and the people.

> ## Smart Cities
>
> *a vision of what is desired and the political will to accomplish it*

Not only does there need to be a strategic plan to turn a city into a Smart City, but the Smart City vision and execution blueprint also need to fit and interact with an overall urbanization plan. Urban planning is concerned with city layout, physical infrastructure, land use, and the use of resources such as water, parks, and conservation of the environment, among other things.

Urban Planning is a full discipline *per se,* and we must not forget it when a Smart City initiative takes place. However, is not the intention of this book to explore that meaningful relationship, as we consider that a Master Urban Plan needs to exist as a foundation for a Smart City initiative or we need to include the development of an Urban Planning phase along with the Smart City Initiative.

4.1   The Framework

As mentioned above, the strategic plan is essential to set goals and objectives and design the programs to achieve them. It also has to identify sources of funding to assure that sufficient resources will be available to carry out vital programs and projects. There are many to consider, and these elements need to be integrated into a framework to include the necessary infrastructure, the information systems, the organizational model, and the required skilled people. In Figure 9, we present a high-level vision of the elements in our proposed framework.

Figure 9: Smart City Framework

Source: Authors

4.2   Vision/Political Will

It is of utmost importance for the municipal government to have a vision of what it wishes to accomplish, as well as the leadership to implement the programs needed to realize the vision. Many management schools, successful leaders, and international political organization support the concept of a broad view to drive the ambitions and guide the behavior of citizens and related stakeholders. For Instance, both the OECD and the European Union have set forth sound principles and guidance for their member states where Vision, Political Will, and Governance are vital components. On the other hand, the accumulated experience(s) from many eGov and Smart City implementation cases indicates that the lack of vision usually produces poor results. An interesting example is the case of Hong Kong, that shows how a lack of vision can negatively impact the outcomes of a project.

4.2.1   Hong Kong Renews Smart City Efforts

Hong Kong, a Special Administrative District of China, has a population of over 7.4 million. It is the seventh-largest trading entity in the world and enjoys a GDP per capita of over $48,000. Its physical infrastructure gives it the position of the fifth busiest container port in the world and the busiest airport for international cargo.

At the end of 2017, Hong Kong issued a Smart City Blueprint.[27] At that time, however, Hong Kong was a laggard in Asia concerning Smart City metrics, and the lack of leadership and vision on the part of the government was the perceived obstacle. Evidence cited included the fact that few people in Hong Kong were using e-wallet technology to pay at restaurants and taxis, whereas in Singapore and Shanghai, the practice was almost universal. A 2017 Smart Cities Index published by EasyPark[28] found that Hong Kong performed poorly in several areas, including transportation/mobility, digitization, and innovative economy. It ranked behind Tokyo, Seoul, Osaka, Taipei, and even Daejeon, South Korea's fifth-largest city. One observer pointed out that "the government is not even tech-savvy itself,"[29] with the highly bureaucratic administration still relying mainly on paperwork. Further, past attempts to build a Smart City had been done piecemeal and lacked an overall strategy.

The current Hong Kong Smart City Blueprint seeks to make use of innovation and technology to address urban challenges, enhance the effectiveness of city management, and improve the quality of life, as well as to enhance Hong Kong's attractiveness to global business and foster continued and sustainable economic development. There are plans and strategies for various aspects such as mobility and environment, among others.

The government's goals and commitment are welcome; however, one analyst believes that "the government's grand plan to promote innovation won't get far without efforts to accelerate digitization in SMEs, strengthen e-government services, and popularize STEM education in schools—in other words, make technology accessible to the masses.[30]

4.2.2 The OECD Principles

The Organization for Economic Cooperation and Development (OECD) articulates its main guiding principles for e-government,[31] and three of their pillars (as seen in Figure 10) refer to Political Will, Vision, Strategy, and Governance. Although the document does not explicitly mention Smart Cities, we see no obstacles, why not the same principles could not be equally applicable to Smart City programs.

Figure 10: OECD Guiding Principles for e-Government

Principle 4 (Vision & Strategy)

- **Adopt a government-wide digital government strategy**

Principle 5 (Political Will)

- **Ensure leadership and political commitment**

Principle 6 (Governance)

- **Establish effective organisational and governance frameworks**

Above all, the OECD emphasizes that e-government is not a goal in itself; e-government serves as an enabler and should be part of a broader policy and service delivery goals. Further, it is most effective when government entities work together. The OECD considers that ICT spending should be an investment, with an expected return.

It should be noted, however, that the return on investment (ROI) for a government program or project is not the same as the ROI for a business venture. A business venture in the private sector measures ROI in terms of profits.

Government undertakings do not necessarily expect profits, and they may not even generate revenue. Nevertheless, an ROI can be in the form of cost savings or broader social benefits. Governments can achieve efficiencies from improvements in productivity, quality, or time savings, as well as direct cost reduction. Such savings and social benefits can be measured.

Phillips[32] of the ROI Institute describes the ROI Methodology for public programs as a cost/benefit analysis in five steps:

- Identify program benefits.
- Convert benefit to monetary value.
- Tabulate fully-loaded costs of the program.
- Identify intangible benefits.
- Compare the financial benefits to the costs.

Thus, society can expect accountability from public sector programs, and using non-economic metrics in the evaluation of Smart Cities and other governments projects is a growing trend (see Chapter 7)

4.2.3   The EU's Digital Single Market Vision

An excellent example of how political will can work for the benefit of deploying eGov solutions (and be useful for Smart City initiatives) is what the European Union is doing regarding policies.

The EU is in the process of a digital transformation of government, and this transformation is one of the significant elements of achieving their Digital Single Market vision and EU2020 goals. A Ministerial Declaration makes explicit the drive towards:

> **...open, efficient and inclusive, providing borderless, interoperable, personalized, user-friendly, end-to-end digital public services to all citizens and businesses—at all levels of public administration.**[33]

We summarized the main metrics that the EU uses to measure its progress in e-Government in Figure 11:

Figure 11: EU Metrics for e-Government

Key enablers:

- refer to the extent to which technical pre-conditions for e-Government service provision are in place. At the national level, this may include e-ID, e-Documents, and Authentication Sources. At the municipal, city level, it involves things like building permits and vendor registrations, many of which can be obtained online nowadays.

User centricity:

- refers to the extent to which a service, or information concerning the service, is available online. By this measure, e-Government in the EU continues to progress, with more and more online services available, and with service quality improving. Further, more than half of online services are available on smartphones or tablets.

Transparency:

- refers to the process of service delivery, the government's responsibilities and performance, and the personal data involved. This area has seen modest progress.

Cross-border mobility

- is where users can use online services in another European country. Although cross-border mobility is improving for both citizens and business, e-government services remain friendlier to domestic users than to users from other EU countries. Cities may not deal with borders with other countries, but many metropolitan areas are composed of multiple municipalities, and cooperation is vital for programs to be successful. However, we need to consider that the concept of mobility goes beyond transportation, as previously addressed on the Smart Territory topic.

4.3   Strategic Plan

The political leadership needs to lay out a strategic plan to express its vision and set the foundations for the Smart City project execution. The strategic plan should contain all the required elements to transform the vision into reality, such as the goals, objectives, metrics and targeted KPIs (Key Performance Indicators); it also needs to identify the policy aspects and associated risks and describe the organization and related stakeholders. A strategic plan should also indicate the management roles and how the city will provide for all necessary resources.

The strategic plan serves as the overall blueprint that will guide all city actions toward the conceived vision. It is the crucial element for guiding the governance bodies and the decision-making processes during all phases of implementation, to assure that the decisions to allocate and prioritize resources according to the guidance or vision of the strategic plan.

4.4   Tactics

In formulating the strategy, we set big plans and bold goals to achieve the desired results. In tactics, we begin to define the specific goals and the means to ensure the execution of the project according to the defined strategy. In a public sector context, this process is done mainly through the public policy process, as we will see next. The specific tasks, initiatives, programs, and KPIs will be set by the proper policies or mandates, and the governance and management bodies will decide and be responsible for execution activities.

4.4.1   Policies

To implement public policies, governments create the laws and regulations, which are the instruments to realize political will. Public policies can also include strategic plans to guide cities in their efforts to achieve operational efficiency with a more significant social and economic development, serve the citizens, and attract business and industry. Policies need to be explicit regarding the way that agencies, citizens, and partners interact. Such policies, in the form of a guidebook, will put the framework into context, describe the technical content, specify process documentation needed, and set forth implementation and compliance regimes.

4.4.2   Governance

The delivery of e-Government services or Smart City initiatives will always face complex management decisions. The allocation of available and usually limited resources tends to generate conflicts among the different stakeholders. Which projects should we fund, and which ones should we postpone or cancel? These questions cannot be answered solely by one individual or a single management team; this is why it is necessary to have a governance process. As stated by Viscusi et al.,

> The focus on back-office improvement is considered as the strategic way to create more efficient and customer-centric public services, where the challenge for public administration is to align IT and organization in order to ensure that their teams can exploit systems to deliver value-added services,[34]

To make decisions following the maximum interest of all stakeholders and within the strategic plan, we need a governance process to guide the organizations leading the Smart City initiative.

Governance encompasses the appropriate decision-making rules and procedures needed to ensure that investments in resources align with the priorities and goals defined in the planning documents for the Smart City initiative. It is also designed to ensure that the decisions are compliant with the social, regulatory, and market environment, including the resolution of conflicts of interest.

There are many methodologies we can apply to ensure the achievement of these stages. Some cities may adopt COBIT/ITIL for IT governance and services, Scrum or Agile for system development or adopt best practices from project management technics. A brief description of these methodologies is summarized in Figure 12.

Figure 12: Governance and Project Management Methodologies

COBIT
•Control Objectives for Information and Related Technology is a framework created by ISACA for information technology (IT) management and IT governance. It is a supporting toolset that allows managers to bridge the gap between control requirements, technical issues and business risks.

ITIL
•Formerly an acronym for Information Technology Infrastructure Library, is a set of practices for IT service management (ITSM) that focuses on aligning IT services with the needs of business.

AGILE
•Is a set of principles for software development in which requirements and solutions evolve through collaboration between self-organizing, cross-functional teams. It promotes adaptive planning, evolutionary development, early delivery, and continuous improvement, and it encourages rapid and flexible response to change.

SCRUM
•Is an iterative and incremental agile software development framework for managing product development.

Although the methodologies above are more related to IT projects, we can apply the concepts to other types of projects, mainly those for Smart Cities where the ICT components represent a big piece of the puzzle. The critical point is to make sure that a formal process is selected and used to govern the teams involved in the Smart City initiatives. In addition to that, part of the governance and management processes requires the adoption of KPIs, not only in the deployment phase but in the ongoing operation phase. We will discuss some of these KPIs in a later chapter.

Assigning responsibility to a lead agency or project office to provide an institutional base for the Smart City effort is also considered a good practice. To carry out its duties, the organization needs to have sufficient resources.

The responsible entity develops a plan of action and assigns tasks to workgroups, coordinates with the various stakeholders involved, and prepares documentation. Other agencies may provide inputs, support, and help with resources and operations. In some instances, incentives may be devised to encourage participation. Also, government personnel needs training on the framework and standards. Cross agency collaboration is usually a complicated topic that needs to be faced and managed by higher levels within the hierarchy in the city. There must be political will and strong leadership to overcome the issues arising from collaboration needs.

It is also a good idea to establish mechanisms, such as public hearings, requests for comments, and Websites, through which citizens can provide input from their perspective. It usually takes multiple iterations of drafting, doing pilot projects, and implementing. Over time technologies change, processes change, and standards may become obsolete. Therefore, the framework should be reviewed regularly and revised/updated when necessary[35]

Planning, implementing, and managing a Smart City calls for new governance models. It is essential to break down the "silos" of different government agencies. In conventional city operations, each department usually has its own IT island; and they do not communicate with each other. Exactly as required by a successful eGov implementation, a Smart City deployment calls for sharing of information among departments and offices.

EXAMPLE: BULGARIA

An example of a country making good use of its limited resources to provide information to the citizenry is Bulgaria. This country's formulation of a strategy also applies to a Smart City effort. Bulgaria addressed the different aspects of the undertaking, including organization and coordination, the regulatory framework, and the resources required.

Bulgaria, a lower-middle-income country, jumped 79 places from the 2014 UN e-Government Development Index (EGDI) survey to a score of 0.6376 and took the number 43 spot in the EGDI Top 50.[36] One factor in Bulgaria's rapid progress is that it has aligned its e-Government strategy with the Digital Agenda for Europe.

In its e-Governance Development Strategy for 2014-2020,[37] Bulgaria analyzed its efforts, which began in about 2001, and they found that a lack of sufficient coordination between the general policy on providing administrative services and the services by electronic means harmed e-Governance development.

The existing regulatory framework was able to accommodate e-Government, but it did not stimulate e-government development. Also, legal problems arose regarding the use of e-documents and e-signatures. There was a significant problem with achieving interoperability, and it was frequently impossible to attain high efficiency, due to a lack of a systematic approach to e-Government development.

The vision for 2020 included the following strategic objectives:

- Provide qualitative, efficient and easily accessible e-services to citizens and businesses
- Transform institutions into digital administration by integrating information processes
- Promote access and participation.[38]

Its e-Governance model is in Figure 13.

Figure 13: Bulgaria's e-Governance Model for its 2014-2020 Vision

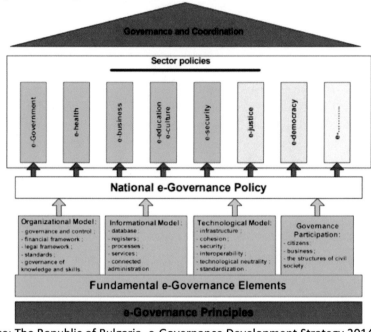

Source: The Republic of Bulgaria, e-Governance Development Strategy 2014-2020 in the Republic of Bulgaria.

The government assessed the resources necessary to achieve the objectives and examined funding mechanisms for the long-term.

An EU report summarizes Bulgaria's progress in e-Government.[39] At the end of 2015, the Ministry of Transport, Information, Technology, and Communication (MTITC) completed its e-Government system, part of an end-to-end solution to provide e-Government services. It also launched a process of implementing the delivery system in all public services.

During 2015, work accomplished includes, among other activities:

- High-speed broadband services were made available to citizens and businesses in economically disadvantaged and outermost regions in the country.
- The government launched a government cloud service platform for all of the country's municipalities.
- The first datasets on its open data portal were published.

The main components of Bulgaria's e-Government infrastructure include portals, networks, eIdentification/eAuthentication, and knowledge management. The e-Government provides a one-stop-shop with a repository of public services provided by the central state administration. Citizens, and businesses can obtain information online on a variety of public services, as well as forms to download. Some 1,300 services are available in various ministries, agencies, and municipalities for matters such as health, vehicles, work/retirement and education for citizens, and startup, customs/VAT, public contracts and product requirements for businesses.

The legal framework includes legislation on e-Government, freedom of information, data protection/privacy, e-commerce, e-communications, and e-procurement.

In early 2017, the State e-Government Agency of Bulgaria officially launched its website, E-gov.bg, to take over the management of state registers and services.

EXAMPLE: CITY OF BOSTON'S DIGITAL GOVERNANCE

In 2010, the City of Boston rebranded its MIS Department as the Department of Innovation and Technology (DoIT). This new department seeks to integrate technology across the organization. At the same time, they formed the Mayor's Office of New Urban Mechanics (MONUM) with a team dedicated to "build a bridge between innovators inside of City Hall and those throughout the community"[40] therefore enabling the DoIT to take its IT agenda to a higher level.

The City seeks an efficient government that empowers constituents and enables connections with and between the city's constituents—public, private and academic organizations, neighboring towns, and peer cities world-wide to formulate innovative solutions to the challenges that Boston and other cities face. DoIT's digital strategic plan is to determine how information and technology are used to reach these goals.

DoIT's principal elements are summarized in Figure 14:

Figure 14: DoIT's Principal Elements

Core Infrastructure

- Technology infrastructure is essential to the delivery of government services. In the 21st century; the "technology stack"—devices, connectivity, servers, and software—as essential to government operation as electricity and roads.

Tools for Government

- City departments demand better tools and technologies, and we must deliver in the same way that we serve our external constituents—by relentlessly focusing on the needs of our "consumers" in City Hall. This empowers the City's workforce to deliver great outcomes for citizens, instead of struggling against outdated process.

Digital Engagement and Service Delivery

- Consumer expectations have shifted as digital technology has transformed the world around us. Government must meet these changing expectations by delivering great service through digital channels. Digital communication technology offers up new channels for citizen engagement and participant in government.

Data and Analytics

- Data driven management helps us to focus time and dollars where they have the biggest citizen impact, improve public accountability and transparency, and measure our progress against broader goals and initiatives.

Broadband and Digital Equity

- Boston seeks to be a city where all residents and businesses have access to affordable, high-speed broadband, and the opportunity to develop digital readiness (connectivity, skills, and equipment, a more connected city is a more equal city, a more innovative city, and a more prosperous city,.

Source: City of Boston[41]

4.5   Resources

When we speak about resources, we can't limit the concept to capital. Resources include elements related to technical, human, and financial components.

Technology is the most visible resource for implementing a Smart City; it provides the building blocks to construct solutions. Technology may be the first resource that people think of, but, technology itself is not enough. Even with a large budget and state-of-the-art technology, a project can still fail. Skilled and adequately trained human resources are vital. However, an overarching requirement is a strategy, a framework, and an implementation plan.

The underlying technology for Smart City projects in sectors such as energy, transportation, and mobility, is composed of sensors, actuators and other CPS (Cyber-Physical Systems), linked to a central operation and control facility through a communications network.

Of particular importance is the IT structure to handle the vast amounts of data generated by the Internet of Things (IoT), and Chapter 5 describes the technology platform components.

However, it is not enough to invest in the technology. Governments have frequently failed with e-government initiatives because they lack expertise in information technology and information systems. The same situation applies to Smart City initiatives.

Financial resources become available through strategic plans, budgeting, financial analysis, and accepted financial management practices and procedures.

Adequate human resources must be available. People designs, implement, operate, and evaluate ICT systems. Managers set objectives, schedules and milestones, manage and oversee the implementation of a project, monitor execution, and evaluate outcomes. Leaders set overall goals, motivate others, and mobilize other resources.

If the necessary qualified personnel are not available, capacity building plans need to be present to assure that these individuals will be adequately trained promptly to perform the technical or managerial duties for which they will be responsible. Therefore, it is necessary to include education authorities among the Smart City stakeholders, to help set the required policies and curricula to assure the proper alignment between what the city needs concerning professionals and workforce preparation.

4.6   Smart City Components

Many of the Smart City solutions can take advantage of existing city infrastructure and can become part of it. For example, networked control for Smart Street Lights piggybacks onto the existing streetlight fixtures. Many other devices, such as cameras, waste collector sensors, or air quality sensors, can connect on the same network hubs.

Various components work together to integrate Smart City solutions into the city infrastructure, including organization and information. The existing databases and other information sources can be available to the public and made searchable through portals and open data initiatives. The municipal organization needs to enable the sharing of information, with each agency playing its part and sharing information.

4.6.1   Infrastructure

A telecommunications network infrastructure is a foundation upon which we can build a Smart City. It must support all applications and services needed for e-Government and Smart City systems such as taxes, permit applications, sharing data across agencies, and citizen services. Smart City solutions can eventually share the same network used by e-Government, but it will probably require a different system, since the IoT devices used for Smart City applications may require different technical specifications or capacities from those needed for e-Government.

The telecom infrastructure is the essential technology for a Smart City and. The other elements of the city critical infrastructure such as bridges, roads, water distributions systems, electrical grid, and hospitals are some examples and will rely on the connectivity provided by the telecom infrastructure to transmit sensor data to a control and management center. Therefore, we need to elaborate proper plans to monitor these elements and assure that they are in acceptable maintenance conditions and that they offer adequate capacity and security to perform reliably and be resilient in disasters and emergencies.

A good example is the Memorial Bridge[42] that links New Hampshire and Maine. Sensors were installed at different points of the span, turning the structure into a "smart" bridge, transmitting information about its stability as well as environmental data on the river.

"We call it a 'living' bridge because it can talk to us and provide valuable information about its health—the stress it deals with, the ease at which it moves, what's happening around it and even under it in the Piscataqua River," said Erin Bell, associate professor of civil engineering and principal investigator of the Living Bridge Project. "This bridge is not just for getting us across the water; it can teach us so much more about the world around us."

4.6.2   Information

The information aspect of e-Government began with schemes for document exchange. Then it progressed to interoperability, open data, text mining, and Big Data. In parallel, more channels have been opened for citizens to communicate and engage with government. The original over-the-counter communication process, to exchange information between citizen and government, has been replaced by call centers, single-window facilities, web interfaces, smartphone applications, social media engagements, and IoT sensing.

This advancement in the ability to obtain data, analyze them, and make them available is a significant element in the evolution of e-Government; and it is one of the key pillars for Smart City solutions. The information generated from these data sources and processes enable governments (including city managers) to understand a situation and make better decisions—and to be proactive rather than reactive.

Combining the availability of real-time data with the analytical insights from the historically collected data, offers the potential to monitor and adjust the delivery of services, enabling improvements to enhance efficiency. Further, where the path of a trend leads to a potential crisis, the government may have the ability to prevent untoward events, rather than be forced to react after they have occurred.

The next phase, with the aid of rule-based systems, machine learning, and artificial intelligence, offers the promise of autonomous decision making in different levels of the government and city operations. In the following chapters, we present the expectations for a Smart City 3.0.

4.6.3 Organization

Planning, implementing, and maintaining a Smart City program calls for an organization. Each participating agency must know how it fits in the overall strategy. The organizational aspect needs to integrate the structure of the organization or program, the systems involved, the workflow, and the culture. There needs to be a shared vision, with leaders able to integrate the work of various groups. The requirements for such an organization include organizational structure, clarity of purpose, and clear roles and responsibilities; and it must adhere to a governance process.

However, creating a new organizational structure can be very difficult because most governments have a very rigid form of hierarchical organization. They usually tend to leverage existing organizational structures, simply assigning them new responsibilities and duties. Many times this engenders conflict with the original or legal attributes of the organization or agency. The result is increased bureaucracy and inefficiency, causing project delays and budget overruns.

Cunningham and Kempling[43] set forth the principles for changing public sector organizations:

- Form a guiding coalition
- Recognize and respond to resistance
- Establish the need for change
- Articulate the expected outcomes
- Establish a process to implement the planned change
- Focus on continuous improvement
- Develop a commitment plan
- Change structures and HR systems
- Managers and leaders involve themselves in face-to-face interactions with staff

A Workforce paper[44] points out that organizational change needs input and involvement from all stakeholders throughout the organization. Further, implementing and managing the amendment calls for proactive communication to inform stakeholders about what is going to happen.

5   SMART CITY TECHNOLOGY PLATFORM

The technology platform starts with architecture. The objective of the architecture is to enable the organization to fulfill the Smart City system design purpose. Achieving this objective requires consideration of all the components of the system and their relationships to each other and the outside world while adhering to the design's guiding principles.

There is a need to design an ITC architecture capable of supporting the broad range of requirements from the various system components that will compose the e-Government services and the Smart City initiatives. This ITC architecture will enable the development and deployment of Smart City Solutions.

Interoperability frameworks were the first attempts to define such an architecture model, resulting in most cases in the integration of systems workflow and a data exchange automata. The usual solution was an Interoperability Gateway based on SOA[45] (Service Oriented Architecture) including a Message Communication Bus. That architecture became the typical pathway for advanced and integrated eGov deployments, and several industry vendors broadly promoted it as the solution for the issues faced by governments trying to implement an Integrated and interoperable eGov architecture. This architecture was instrumental in supporting the initial requirements of an integrated eGov where most of the systems need only to exchange messages across structured data and were limited to addressing a very well-defined set of processes, leaving very little room to deal with exceptions and new requirements. That stage became eGov 1.0.

As we moved eGov into 2.0 and 3.0 levels, new technological solutions were necessary to support the complex set of logical process and data volume requirements. The development of applications to address 2.0 and 3.0 implies an exponential complexity of the algorithms and methods to deploy the necessary solutions. When this happens, there is an increased need for computing power and data-storage capacity, as well as a more comprehensive and robust telecommunication infrastructure.

Some of the functionalities and services of eGov are an integral part of a Smart City's existing eGov platforms; if deployed in a City, they will need to interoperate with, or be part of, a broad Smart City Platform. To that extent, we will now point out the essential needs for such a Smart City platform; in particular,it needs to coexist with other government systems and solutions already available in the city and, at the same time, integrate new technologies and provide a full range of new services. Figure 15 depicts a generic architecture diagram indicating the main components necessary for a Smart City Platform.

Figure 15: Generic Architecture for Smart City Platform

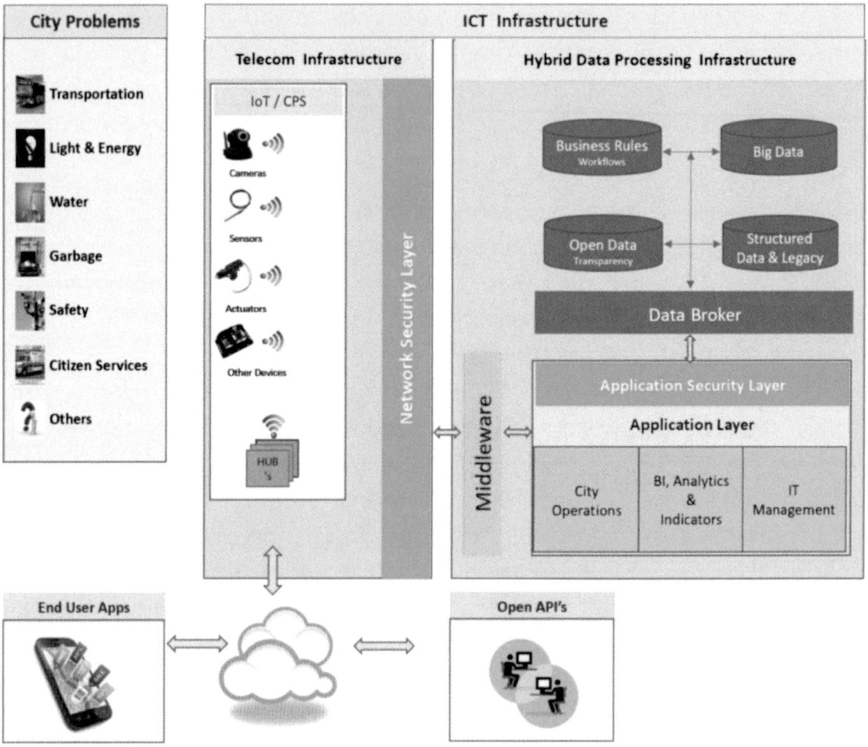

Source: Authors

5.1   Architecture

The architecture integrates various elements with a foundation based upon an ICT infrastructure, where the multiple components can connect to exchange information. There are hardware and software elements, such as the IoT devices, CPS, telecommunication networks, databases, middleware, and a variety of applications to provide information to citizens, city managers, and high-level decision-makers. It integrates and interoperates with eGov systems by sharing structured and open data and also provides Citizen Services.

Under this model, city managers can interact with the City infrastructure. For example, IoT devices connected to the platform can collect traffic data, analyze its patterns, and send commands back to traffic lights to optimize the flow on a congested road. Such a platform needs to isolate the different functional layers to be as flexible as possible to allow adding new components seamlessly. In this example, we list the main platform components in Figure 16:

Figure 16: Main Technology Platform Components

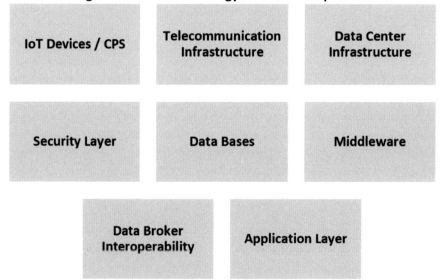

The application layer will have to support several user requirements, as seen in Figure 17.

Figure 17: User Requirements supported by the Application Layer

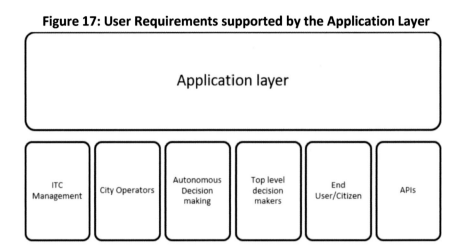

Let's discuss some of these modules or components

5.2   IoT (Internet of Things) Devices /CPS

Current hardware technology is creating hundreds of new devices that can sense the state of almost any asset in a city and, using the IP protocol, transmit this information to a control system. The system can then act upon the data received or store it in a database for further statistical analysis and pattern identification. The IoT devices are not limited to sensors, but they can include other functional devices such as cameras, actuators, or even other fully autonomous cyber-physical systems (CPS). The commonality is their capability to transmit and receive information using the IP protocol.

The Smart City Platform needs to have the capability to accept any IoT device or CPS, regardless of the function it performs. Therefore, when we refer to IoT devices, we also include CPS. Software drivers must be provided by the vendors, to ensure compatibility and support to different API protocols. A middleware layer, as described in its section below, will handle and harmonize requests and protocols from each one of the IoT devices, and the target applications.

IoT devices need to be able to push information into the Smart City Platform and its databases; if the device is an actuator, it needs to act and respond upon receiving commands from the Platform.

5.3   Telecommunication Infrastructure

Any Smart City initiative requires an excellent telecommunications infrastructure. Most of the IoT devices will require some form of wireless communication. The list of wireless communication alternatives keeps growing with a forward look to 5G technologies. The need is to remotely communicate IoT devices with their respective software counterparts in the platform. Some cities have decided to build their telecom infrastructure, while others rely on the existing incumbents to provide such service. City managers need to evaluate the best alternative that fits their operational and strategic needs as well as their budgets.

The usual Smart City telecom network aggregates and combines distinct technologies to provide city-wide coverage: Fiber optics, cellular networks (3G/4G/LTE), Wi-Fi, LoRa, ZigBee, and other proprietary communication networks. The decision as to what technology to use depends upon several factors, such as the application bandwidth requirements, device power consumption, reliability, CapEx/OpEx, and even the business model associated with the network. Cities can own their telecom network or exchange right-of-way with telecom incumbents for fiber optics access. Therefore, the city should have enough technical knowledge among its ranks to make an educated decision when selecting a telecom network solution. The use of external consultants is a good alternative when that capacity is not available within the city government personnel.

5.4   Data Processing Infrastructure

The data center serves as the primary installation for data storage and processing for municipal institutions.

City Managers responsible for implementing this framework will have to decide which is the most convenient data processing infrastructure to support the deployment of the Smart City Platform. The options will range from building and operating their Data Centers to utilizing resources provided by cloud computing vendors.

The current trend toward cloud computing involves replacing servers and computer/storage resources with outsourced services. Cloud computing enables users to increase capacity or add capabilities without investing in new infrastructure, training new personnel, or licensing new software. The adoption of cloud computing offers many benefits; however, most of the existing e-Government solutions still run on traditional in-house data centers. Therefore, transitioning to cloud computing may be cumbersome, expensive, and a time-consuming process that will be achieved only over some time. For that reason, the team responsible for defining the data processing infrastructure for their Smart City implementation will consider running the Smart City platform using a hybrid model, to assure that on-premises applications can be seen and can interact with other cloud components in the platform. At the same time, it is necessary to ensure portability to the cloud of their existing on-premises applications, with minimal operational impact.

The current service models for cloud computing adoption are Infrastructure as a Service (IaaS), Platform as a Service (PaaS), and Software as a Service (SaaS).

The IaaS model is the most basic and offers virtual computing resources such as CPUs, operating system images, raw and file-based storage, firewalls, load balancers, IP addresses, and other elements. These resources become available through the internet or private networks. Large equipment pools installed in data centers will feed the virtual requirements.

In the PaaS model, a computing platform is provided, typically including operating system, programming language, execution environment, database, and the webservers. Application developers can develop and run software solutions on this platform without the cost and complexity of buying and managing the underlying hardware and software layers. The computer and storage resources scale automatically to match application demand so that the cloud users do not have to allocate resources manually.

In the SaaS model, the provider installs and operates application software in the cloud, and cloud users access the software through the internet, therefore eliminating the need to install and run the applications at the user's premises, which simplifies maintenance and support. This model enables scalability by cloning tasks onto multiple virtual machines to meet demand. Load balancers distribute the work over the set of virtual machines.

Frequently, city managers are positioned to consider a binary decision between adopting an on-premises or a cloud computing solution. However, the alternatives for a hybrid solution are gaining traction, and the leading vendors in the market are offering innovative technologies, making it easier to move data processing workloads back and forth from an on-premises environment to a public cloud. A decision on what to choose will consider cost optimization, security, and reliability, as well as technical factors such as latency, bandwidth availability, and business continuity.

Among the hybrid alternatives, the use of a vendor's data center to host the city ICT infrastructure is also a good alternative.

5.5   Security Layer

The platform needs to support different security components. Security layers should have at least two main parts, one related to the telecom network, and a second, related to application utilization security.

Encryption, single sign-on through federation services, digital certificates, and digital signatures are some of the security elements we need to implement in the platform. It is not our objective to detail these components, but alert the team deploying the platform infrastructure about the risks of inadequate preventive planning.

Another topic of security is related to the tools and mechanisms to support citizen privacy regulations, such as the General Data Protection Regulation (EU) 2016/679 ("GDPR")[46.]. This is a regulation in EU law on data protection and privacy for all individuals within the European Union and the European Economic Area (EEA). It also addresses the export of personal data outside the EU and EEA areas. Similar regulations are already appearing in other geographical regions, and the security levels in a platform will need to withstand the requirements for such type of data protection.

5.6 DataBases

A Smart City technology platform needs to support different kinds of database structures. There are, however, four primary classifications for the type of data they contain. Table 1 presents some of these database components.

5.1 Middleware - Data Broker – Application Layer

As defined by Microsoft, middleware is software that lies between an operating system and the applications running on it. It functions as a hidden translation layer, enabling communication and data management for distributed applications.[47] Some authors describe middleware as "software glue."

Middleware makes it easier for software developers to perform communications across different IoT devices so that they can focus on the specific purpose of their application. Middleware is the software that connects software components or applications to other software layers. Middleware is the software layer that lies between the operating system and the applications on each side of a distributed computer network. Typically, it supports complex, distributed business software applications.

Table 1: Database Components

Type	Structure	Access	Used by	Usual Platforms
Legacy Data	Structured	Authorized systems only	Legacy Systems; Management Information Systems	SQL, Oracle, DB2 & other DBMS Products
Open Data	Semi-Structured	Open, Publicly available	any system or citizen	SOCRATA, CKAM, Open Standards Files & others.
Big Data;	Structured and Unstructured	Authorized systems only	Security, Planning, and analytic applications	NoSQL DB such as Mongo DB, Cassandra, MarkLogic & others
Business Rules	Structured and Unstructured	Authorized systems only	BPM Suites, BPM Applications, Middleware Systems	NoSQL and SQL DBs

Source: Authors

This definition applies perfectly to the functions we expect from the middleware layer in the Smart City platform. Since there will be thousands of sensors collecting billions of data points in a Smart City, the Middleware needs to create an abstract logical layer to enable the systems running in the applications layer to seamlessly communicate with the sensors, regardless of make or model or protocol used by any device. The middleware is, therefore, a key component to this architecture.

Side by side with the middleware, a data broker component could facilitate the exchange of information among the different databases in the platform. Like a communication bus in an SOA (Software Oriented Architecture) model, the data broker will exchange information across the platform, providing data elements to any user, once the requestor had received clearance by the business rules controller.

5.2   Application Areas

The application areas cover those activities that can be made more efficient by technology, such as transportation and traffic, energy and environment, safety and security, citizen and business services, city infrastructure operation and management, and internal government applications.

The goal for Smart City initiatives is to improve the management of public services allowing for real-time responses to incidents. In the Application Matrix in Table 2, we describe the categories for Smart Cities applications (including the ones typically considered eGov applications) not only by their areas of concern but also by the different audiences they will serve. We aggregated these audiences into three groups: decision/policymakers, city managers/ operators, and end-users/citizens. On the Appendix of this book, there is a comprehensive taxonomy to categorize Smart City initiatives.

Table 2: Application Matrix Examples

Audiences →		Decision/Policy Makers	City Managers/ Operators	End-users citizens
System Category →		DSS Decision Support Systems	MIS Management Information Systems	Apps Applications for citizens
Service Vertical	Transportation and Traffic	Origin-destination systems: Where/when to build a new road?	Monitor transit to decide daily operations: # of buses. Train frequency, routes, others.	What bus do I need to take from A to B? When is the next train arriving?
	Energy and Environment	Weather, flood impact simulation Waste collection optimization	Real-time air/water quality monitoring system.	Air/water contamination Info & Weather Alerts
	Safety and Security	Crime Heat Maps	Theft alarms Camera surveillance	Red, Yellow and Amber Alerts
	Citizen and Business Services	Citizen Sentiment	Permits, licenses	Library, tourism, transportation, city info
	City Infrastructure	Planning and simulations tools for traffic, zoning, events	Mass transit & road operation	Online information about facilities, hours of services and events
	Government Applications	Budget/Forecast	Tax, registers, ledgers	Pay fees Communicate issues Request services

Source: Authors

5.3 Other Architectures

In the previous section, we proposed and described a base architecture, which illustrates the main components required for a Smart City Architecture, and we also wanted to show the complexity of such requirements. The previous diagrams and descriptions do not intend to be a complete representation of implementation needs.

Figure Figure 18 shows a more detailed architecture for a Smart City Platform, as proposed by the Government of Catalonia in Spain. It provides additional components, using a slightly different taxonomy, but the objective is similar: To structure the requirements and functionalities for a sophisticated platform.

The objective of the architecture is to enable the organization to smoothly create applications and systems to fulfill the purpose of their vision. Achieving this objective requires consideration of all system components and their relationships to each other and the outside world while adhering to the design guiding principles.

Nevertheless, the complexity of such architectures will probably require different levels of specialization, and in some cases, we will see the need for specific sub-architectures to deal with some macro components or CPS. For instance, the NSIT developed a framework[48] just for Cyber-Physical Systems (IoT), the CPS PWG Cyber-Physical Systems (CPS) Framework Release 1.0, as depicted in Figure 19.

Figure 18: Detailed Smart City Platform for Catalonia, Spain

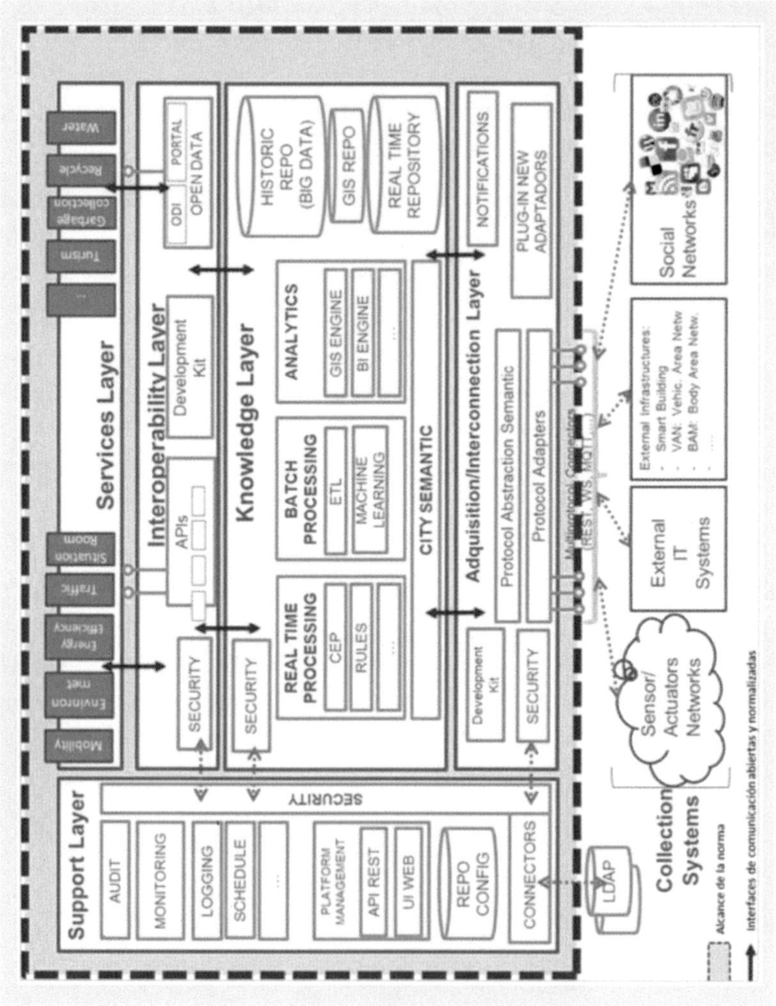

Source: https://enginyeriainformatica.cat/?p=22631

Figure 19: NIST Framework

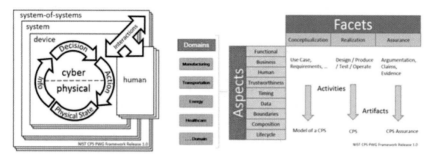

CPS Conceptual Model   CPS Framework – Domains, Facets, Aspects

Source: NIST.

Another case is the NIST Big Data Reference Architecture[49] that further assigns organizational roles and responsibilities to the traditional interoperability architectures to support a Big Data program.

Those roles include:

- Data Provider (including IoT components)
- Big Data Application Provider
- Big Data Framework Provider
- System Orchestrator
- Data Consumer

Additional functional responsibilities include:

- Security
- Privacy
- Management

The Architecture diagram is in Figure 20

Figure 20: Architecture Diagram

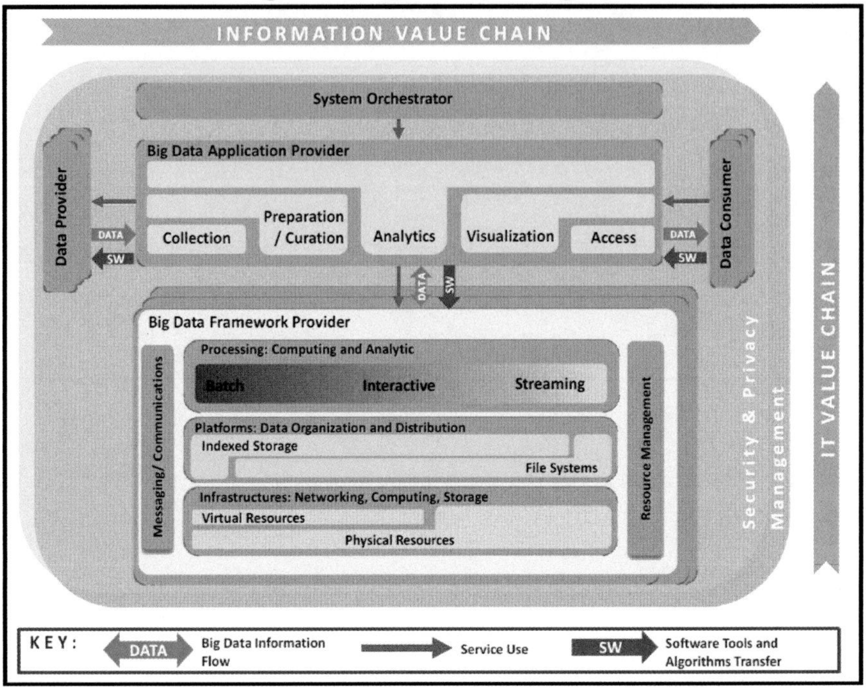

Source: NIST

This NIST architecture provides a framework to support a variety of business and government environments, including tightly integrated enterprise systems and loosely coupled vertical industries, by enhancing understanding of how Big Data complements and differs from existing analytics, business intelligence, databases, and other methods.

The document "A Consensus Framework for Smart City Architectures."[50] discusses another significant effort to provide guidance and structure toward Smart City deployments. Released in September 2018, with considerations for an Internet-of-Things-Enabled Smart City Framework. The document first paragraph states:

> This IES-City Framework is the product of an open, international public working group seeking to reduce the high cost of application integration through technical analyses of existing smart city applications and architectures.

60

This IES-framework is not an architecture model, but rather, a set of tools and principles, *"intended to allow city stakeholders of all kinds to reduce the barriers to deploying technology innovations and enhancements to city life."*

5.4   New Technology Trends

"Regulations moves with the speed of law;
Technology changes with the speed of light."

We regularly use the above quotation in our presentations. Although anecdotal, it represents quite well how the regulatory pace cannot cope with the speed of technology changes. Another quote, which also expresses a similar concept, comes from Arthur C. Clarke: "Any sufficiently advanced technology is indistinguishable from magic."

Given the fast pace in technology change, it is not unusual that an adopted solution becomes obsolete in a couple of years, and sometimes even within months after being introduced. Therefore, trying to forecast what lies ahead in this area is a challenging exercise, and we should leave it to futurists and specialized researchers.

Nevertheless, we need to point out two specific trends in technology that may substantially impact how we plan for a Smart City in the upcoming years: The use of artificial intelligence for decision making and the Blockchain technology.

5.4.1   Using AI for Decision Making

An essential element in government administration is decision making. Large and small decisions are made every day at all levels of government. For example, an official at the social benefits agency approves or disapproves a disability claim; a mid-level official selects a new school computer system among competing bids, or the head of an agency adopts a new policy. All these activities will directly or indirectly impact the citizenry.

As Pearson notes,[51] public administrators typically make decisions through a two-step process. First, they need to determine the requirements of the applicable law. If the law allows them to use their discretion, they generally make decisions like most people: considering the facts and the relevant values, they seek to maximize utility. However, for most routine government decisions, the law does not provide for discretion. For example, the establishment of social benefit amounts for an individual uses complex formulas emanating from the legislation.

The continued development of technology is expected to lead to the use of autonomous decision making, based on Artificial Intelligence (AI). AI uses software to imitate intelligent human behavior. AI programs are designed to learn, reason, and make decisions; they can carry out activities such as monitoring, discovering, predicting, and interpreting. AI is already superior to humans in skills such as playing chess. AI programs can read X-rays and other forms of medical data and can detect heart disease and various types of cancer more accurately than humans. AI systems also carry out functions such as inspecting bridges. There is no reason why we should not apply these advancements to e-Government systems and Smart Cities: human reasoning is combined with machine interpretation, automating most of the task. For instance, a smart camera can, without human intervention, identify a car driving the wrong way; the system immediately triggers safety protocols to minimize or avoid the risk of an accident, either by automatically dispatching the traffic police to the area or closing the access to that road by manipulating traffic signs.

Computer reasoning based on logic and deduction, optimization, and decision making enables autonomous systems and decision support aids. Although AI offers substantial benefits for government operations, there is a high failure rate when governments attempt to adopt it for government purposes. Structured participatory processes are available, but government entities seldom use them in decision making. Participation and Consensus aspect are fundamental when developing autonomous decision-making systems.

As Sundberg and Larsson[52] note, successful decision making needs a formal process involving several factors, summarized in Figure 21,

Figure 21: Factors Involved in Decision Making

Defining and weighing objectives	Analyzing stakeholders	Planning activities and allocating resources
Identifying and assessing risk	Setting indicators for measuring outcomes	Determining if objectives were met, as input to subsequent decisions

A structured process is even more critical when designing a system to make autonomous decisions

Further, in e-Government+Smart City, technological and political factors, organizational and institutional factors, and legal and regulatory factors need to be considered. Decision making becomes very complicated when multiple systems and stakeholders need to collaborate and interoperate. Rational decision making, based on a structured process is more likely to yield the desired outcome.

A Deloitte study[53] finds that if a government formulates metrics to measure the results of its operations, then it can employ data-driven analytics to make better decisions and allocate resources for optimal results.

The public safety grants at the U.S. Department of Justice's (DoJ) Office of Justice Programs (OJP) provide an example. For many years, during the summer grant "season," around $2 billion would be distributed to some 2,000 grantees. In the past, there was no standard approach for determining who received grants. The individual grant managers used their judgment, basing their decisions to a great extent on their knowledge of the applicants.

Then, about 2011, OJP began introducing objective measures into the grant review process and then automated the process. The new system resulted in increased accuracy and consistency of decisions, as well as a more efficient review process. The time for a grant manager to capture grantee data in the database decreased from 30 minutes to almost zero, and grant applications could be reviewed quarterly instead of annually. OJP now allocates its resources based on hard data rather than subjective opinion.

Efforts are ongoing throughout the government to achieve more scientific management and replace intuition with objectivity. This effort emanates from the 1993 Government Performance and Results Act (GPRA), which requires federal agencies to include performance management as part of their strategic planning.

The Deloitte study sets forth four steps to analyze a mission to enable autonomous decision making, summarized in Figure 22.

Figure 22: Steps to Analyze a Mission

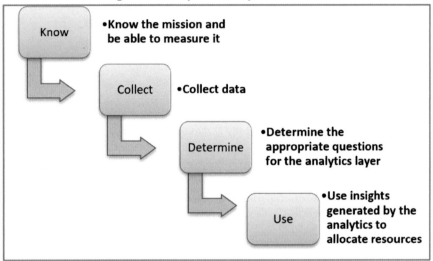

The first step is to find ways to quantify the mission. Metrics may measure inputs, outputs, or outcomes. Inputs are funding, personnel hours, and other economic factors. Outputs are government management KPIs, such as the number of homes visited by social caseworkers. The most important thing, however, is the outcome—the consequences of the program that directly impact the citizens. The result may be, for example, the number of homeless individuals who have gone through occupational training, found a job, and now live independently. Then the agency needs to create a platform designed to collect, store, and disseminate the relevant data, which may mean going outside of the agency to obtain information from other entities. Therefore it is necessary to break down the silos. Third, the use of analytics enables the government to derive meaning and insights from the data it collects. Finally, in Step 4, data insights obtained from the operation can be translated into organizational action to implement operational changes.

5.4.2   Blockchain

Blockchain is a list of records (called blocks) in sequence. Each block contains a timestamp, a cryptographic hash of the previous block, and transaction data. Each record links to the one before it, hence the "chain." The cryptographic hash of the previous block combined with the iterative process confirms the integrity of that block, going back to the original block. We can use Blockchain as a distributed ledger managed by a peer-to-peer network with a protocol for communication and validating new blocks.

The characteristics of blockchain make it resistant to alteration. The transactions are verifiable and permanent and thus considered to be secure. Once we register a block, the data cannot be changed retroactively without altering all of the subsequent blocks—which would require the consensus of the majority of the network participants. Further, the peer-to-peer network lacks centralized points of vulnerability and thus has no central point of failure, as each network node has a copy of the blockchain.

Many blockchain applications have the potential to bring benefits to the city if the citizens freely adopt them. On the other hand, there are applications for which the City can enforce adoption to make the City Smarter. Figure 23 summarizes the prime opportunities for blockchain adoption in Smart City projects.

Figure 23: Prime Opportunities for Blockchain

Notary Services	Property Registration (e.g., business, real estate, vehicles)	Smart Contracts between the city and vendors/providers.

For notary services, blockchain is a way to digitize the activities of the notary public to provide a convenient and reliable system to register the transfer of rights and other transactions. The general idea of a notary is "an independent, even-handed witness that documents the presence or absence of a certain fact."[54] For example, a notary may verify the authenticity of a signature or authenticate the correctness of copies of documents. Notarization is intended to deter fraud by assuring the parties of a transaction that a document is authentic and can be trusted.

The notarization process is composed of three parts: vetting, certifying, and record-keeping.[55]

Blockchain technology assures the integrity of data once written in the chain. The promises like tamper resistance, non-repudiation, and its traceability make blockchain a good candidate to provide some of the notarization capabilities.

- KC Tam
Notarization in Blockchain (Part 1)

There are several notary public services already in operation. Currently, such services cannot directly replace existing notary services under the law, but they facilitate or enhance part of the entire process. Further, the law in each country or other jurisdiction will determine whether they are acceptable.

In property registration, blockchain has the potential to enable the transfer of property almost instantly in a secure manner. The ledger would be updated automatically instead of the buyers having to transfer ownership through an application form.[56] The transparency of the blockchain technology makes it ideal for use in public records systems.[57]

In fact, by enabling the parties to timely have the necessary data, all sorts of contracts could be streamlined to be faster and more efficient. Smart contracts could replace many routine actions previously handled by lawyers.

Money transfer, payments, and delivery service tracking for and between third parties are possible services to be implemented by private entities, but usually, not by the city itself. These services are beyond the scope of this book.

6 IMPACTS OF SMART CITY SOLUTIONS

Cities are a significant force in the economy, and they are becoming stronger. Blaž Golob likens modern towns to the Greek city-states.[58] He contends that the national central governments, their political parties, and world bodies such as the United Nations are no longer the main drivers of social and economic development. Cities are a relevant focus of economic activity, and they are rising as new power players. In Europe, for example, 67 percent of Europe's GDP is generated in metropolitan regions, while their population represents only 59 percent of the total European population.

Several international forums have studied the impact of Smart City technologies and find that they have a significant positive effect on development. Through Smart City programs, public administration may be made more efficient and effective. These technologies can dramatically improve the efficiency, availability, and reach of public sector operations.

Table 3 summarizes the impacts on various sectors. The development indicators cover infrastructure and efficiency gains, human capacity building, promoting safety and security, and others, with examples showing how Smart City technologies have provided benefits in other cities.

Numerous cities can demonstrate the positive impact of the development of Smart City programs, and we will present several examples.

Table 3: Development Impact of Smart City Projects

Category	Some Expected Development Impact
Infrastructure Development and Efficiency Gains	• New technologies introduced, increasing efficiency, capacity, or output process improvement • Capacity added, security/redundancy gained, or reliability improved through the implementation of data centers, cloud computing systems, other infrastructure • Enhanced performance of municipal functions (e.g., street lighting, waste management) • Potential to enable innovations that increase efficiency and allow better management of resources
Human Capacity Building	• jobs created by project implementation • Training and skill development during implementation • Development of human resources with Smart City skills, including planning and design, implementation, management, and data literacy
Promoting Safety and Security	• Improved human safety or security resulting from project implementation • Enhanced public safety outcomes by generating public safety data, correlating data sources to create intelligence, delivering intelligence to decision-makers, and reducing response times
Promoting Effective Governance	• Smart City operations centers provide monitoring and management of government services • Enabling and promoting interoperability and sharing of information among departments
Transportation	• Smart traffic lights enable automated sensing and management of traffic to reduce traffic congestion • Smart Parking technology provides real-time information to users
Education	• Remote access to health services and education • Real-time feedback and assessment of student performance • Individualization and personalization of the educational experience
Health Services	• Remote access to health services • Remote diagnosis and treatment • Digital health records • Remote patient monitoring
Power	• Smart Grids • Smart street lighting enables control of each streetlight individually, resulting in substantial savings in electricity
Disaster/Crisis Management	• Identification of risk factors/areas • More rapid coordination and effective response to disasters and accidents

Source: ASTRO Systems.

6.1 City Infrastructure

The necessary infrastructure of a city encompasses housing, sanitation, water supply and sewage, electric power supply and distribution, transportation, waste management, and communication, among others. Adding digital infrastructure with network control to this physical infrastructure provides the foundation for a smart system. For example, the physical infrastructure of a streetlight system consists of the bulbs, poles, and power lines. The digital infrastructure that turns it into a smart streetlight system consists of sensors, controls, network connections, and software. This improved, smart infrastructure allows each streetlight to be operated and controlled separately and to have its consumption measured individually.

ICT infrastructure integrated with these systems can enhance the performance and efficiency of municipal operations by incorporating intelligence into them. These smart infrastructure applications also have the potential to enable innovations that increase efficiency even more and allow better management of resources.

The physical infrastructure for a public transport system can cover both individual mobility and mass transit systems—roads, automobiles, traffic lights, trains, tracks, to name a few. The digital infrastructure incorporates sensors, GPS tracking, dynamic traffic lights, passenger electronic highway signs, automatic license plate readers, signaling systems, and, most important, the capability to integrate live data from these sources. Intelligent Transport Systems (ITS) can lead to improvements in safety, reduction of traffic congestion, improved accessibility, and convenience.

Smart City technology enables cities to make more efficient use of energy and resources. Table 4 summarizes some examples of Smart City infrastructure.

Table 4: Examples of Smart City Infrastructure

System	Example Smart Infrastructure Solution	Application
Buildings	Smart Buildings	Sensors and related technologies to improve energy efficiency, safety, and security
Parking	Smart Parking	Parking lots and street parking locations transmit real-time information to users
Traffic control	Smart Traffic Lights	Automated sensing and management of traffic
Street lighting	Smart Street Lights	Individual control of the street light. Intensity can be adjusted to reduce energy consumption and increase safety
City management	Smart City Operations Center	Monitoring and control of government services such as transport, environmental and emergency services
Health and education	Remote Healthcare and Online Education	Remote access to health services and education

Source: ASTRO Systems.

As indicated in Chapter 5, the infrastructure components consist of IoT devices, telecommunications networks, and data processing sites. On the software side, we have the security layer, databases, middleware, and the data broker. The application layer will use these components to access the data, process it, and provide information to city managers and decision-makers, as well as to citizens. For example,

- In Kansas City, the city is using a 2.2-mile stretch of the downtown streetcar line as the backbone of new digital facilities that include sensor technologies, smart lighting, and digital kiosks. The data obtained by the devices tracks and reduces both water and power consumption and fosters business growth, among other results.
- In Hamburg, Germany, the Hamburg Port Authority needed to improve its traffic control system. Various networks were consolidated to develop a fiber-optic backbone to provide a reliable, secure, and integrated network. A broader network

throughout the port supports some 300 sensors for roadways and incoming ship traffic. This intelligent network helps port operations run smoothly.[59]

6.2   Smart City Operation Center

Most cities looking into Smart City initiatives already have some form of City Operation Center. Usually, we find a couple of stand-alone operation centers that have been created to serve specific city services. It is not unusual that the police department has its operation center, as well as the fire department, health services, traffic operations, water, and other utilities.

As we move into a more efficient and Smarter City, these distributed, vertical oriented operation centers should integrate a single Smart City Operation Center. The benefits are not only due to the economies of scale on the infrastructure cost by leveraging resources, but also by having multidisciplinary teams, from different agencies, working side by side, resulting in a high level of collaboration. A shared operations center among neighboring towns can also benefit from services provided by a regional utility or a central authority, which will collocate their operation personnel in the shared facility

EXAMPLE: RIO DE JANEIRO'S OPERATIONS CENTER

The Operations Center for the Prefecture of Rio de Janeiro brings together the municipality's 30 departments and private-sector suppliers in a single monitoring room. In this facility, personnel track real-time conditions in the city, coordinating a response to emergencies and disruptions when necessary. It is considered one of the most ambitious urban command centers in the world.

A large video wall grid displays status graphs, meteorological reports, and live video feed from traffic and surveillance cameras. The city's information systems are displayed in GIS maps, and the staff can adjust for close-ups and additional data overlays. For example, a screen might show the current location of every city bus, the nearest hospitals, and the route to an emergency facility or designated shelter areas during storms. There is Information sharing in real-time among city staff from various departments, e.g., transportation, sanitation, health, and emergency services, as well as with the private contractors that own the transit lines, perform road work, or collect trash. On an upper level is a room for journalists who can access much of the same information, helping to make this approach more transparent.[60]

6.3   Disaster/Crisis Management

The four primary functions of emergency/disaster management are to:

- Prepare for a potential human-made or natural disaster and to prepare public and private entities to respond in a coordinated, effective manner.
- Respond to the emergency using all the resources available from the public and private sectors.
- Manage the recovery from the event.
- Identify mitigation options that will better prepare the government and the community or prevent a disaster from occurring.

Disaster management is a critical element of a Smart City strategy. Disaster management systems require cross-cutting functions that integrate with transportation, power, security, and health services, among others.

An essential component of disaster management is the effective management of information. Smart City solutions for disaster/crisis management consist of a set of data collection and analysis tools that enable efficient communication among first responders and with the authorities. Emergency management solutions allow city authorities to efficiently collect and aggregate existing and future data, which may be in the form of historical event information, sensor data, system status, and video. It is also essential to be able to share data across organizational boundaries to mitigate disaster or crises.

Sophisticated analytical engines can automatically process available data and provide intelligent insight into key performance indicators and trends. Such solutions may call for the implementation of additional infrastructure in the form of crowdsensing capabilities, video surveillance systems, and other technologies to provide the needed data for emergency management oversight and decision making.

Some of the benefits of Smart City solutions for disaster/crisis management include:

- Social media for crowdsourcing
- Risk-based planning and resourcing
- Supporting sustainability
- Avoiding and thwarting terrorist threats

Social media formats have become a tool for emergency response, as they can be employed to communicate with residents to calm them and assist them in working with emergency responders. Risk-based planning and resourcing enable decision-makers and planners to identify and assess risks and to assign resources accordingly.

For example, as noted in an ASTRO Systems Study,[61] the Chile National Emergency Network incorporates a variety of technologies into a fully redundant, diverse, national communications network encompassing knowledge and experience on disaster crisis management from a smart city perspective.

6.4   Energy

Various operations of a city are substantial consumers of power. For example, streetlights may represent 40%-50% of a city's electricity bill.[62]

Smart City technologies can help reduce these expenditures. Smart energy management systems, employing sensors, advanced meters, renewable energy sources, digital controls, and analytic controls to automate, monitor and optimize energy distribute and use, can help reduce municipal expenditures on power.

The Idea of Smart Energy refers to the idea of meeting energy needs in a cost-effective and environmentally sustainable manner in the long term. Various Smart Energy technologies include Smart Grids, improving the energy efficiency of buildings, and energy storage, among others. The critical element of the Smart Energy infrastructure is the smart grid. A Smart Grid is "an electricity delivery system (from the point of generation to the point of consumption) integrated with communications and information technology for enhanced grid operations, customer services, and environmental benefits."[63] Some of the benefits of a Smart Grid include lower costs, increased system efficiency, and cleaner environment, such as:

- Reduced operational cost
- Increased revenues due to a reduction in energy theft
- Improved cash flow from more efficient management of billing and revenue management
- Reduction in both transmission and distribution losses
- Higher customer satisfaction
- Reduction in peak load and energy consumption
- Improved load forecasting
- Increased capability to integrate renewable resources
- Reduction in carbon emissions resulting from efficient operation
- Lower system losses
- An overall increase in energy conservation

ENERGY EXAMPLE: STREET LIGHTING

Street lighting is an integral part of urban infrastructure; it enhances safety and security in public places.

The new efficient LED lamps for streetlights consume less electricity and enable municipalities to save money. LED lights offer cost savings in both energy consumption and maintenance costs; also, the light that they emit is superior, and the price of the LED bulbs is falling. More recently, networked control technology goes even further by allowing the individual control of each bulb—dimmed or turned off when not needed—which enables even greater efficiencies and cost savings.

Many cities have already deployed LED technology for street lighting, either in pilot programs or full implementation, and they plan to make a complete transition to LED. In particular, the European Union has issued a directive regarding a complete change to LED.

With conventional street lighting technology, all lights are either on or off. However, full lighting may not be necessary all night, or in certain areas for specific periods. For example, when there is no traffic in a given area, the lights do not need to be turned on. If streetlights could be dimmed or turned off when illumination is not required, electricity consumption falls with the consequent reduction in costs.

The transition to LED lights offers a platform for networked control of the streetlights in a city. By connecting LED streetlights to sensors, switches, and dimmers--and linking them to a network (wired or wireless)—the town attains a new level of efficiency. The use of LED lighting alone can offer 50% savings in energy. When combined with a networked control system, energy savings can rise to 80%.[64] Further, these savings accelerate the time to see the return on investment.

6.5   Transportation

Transportation is an essential element in a city, covering both private and public modes of transportation. Transportation enables people to get to work, school, and other activities and allows the delivery of goods and services. One goal of city planners is to facilitate access by all sectors of society, to urban facilities, in all regions of the city.

Some cities emphasize rail systems, with integration and smooth transfer among different transport systems to achieve a coherent overall system. A long-term strategy is designed to ensure the development of public transport infrastructure and services.

Strategies may include, among others:

- Improvement of public transport infrastructure and services, and promotion of public transport
- Enhancement of transport facilities for and encouragement of walking and cycling.

Some objectives regarding public transport may be:

- Expansion of the rail system network and expanding the share of rail systems within public transport in general.
- Development and expansion of the number of transfer centers, improving integration among public transport systems in terms of facilitating the commuter transfer, payment, traffic, and others
- Improve public transport service quality, comfort, and safety.
- Bolster public transport capacity, particularly in busses, in addition to railway systems.
- Prioritize road traffic for public transport, i.e. Metrobus systems.

Not all objectives will apply to every city. For example, a town with no rail system would, of course, not be concerned with expanding a rail system.

TRANSPORTATION EXAMPLE: PARKING

Drivers looking for a parking space are a significant cause of traffic congestion in cities. New Smart Parking solutions can provide drivers with real-time information about available parking. For example, the Spanish company Wellness TechGroup developed a solution using intelligent IP cameras and smartphone apps, that analyzes parking usage and identify vacancy. It can also integrate enforcement applications with expiration notices, which officers can use for more efficient ticketing. The information obtained can also help cities with long-term planning.

The system enables parking enforcement officers to view a live video feed of parking in no-parking zones, helping them to detect and report parking violations. Drivers can search for parking spaces based on their point of interest or preference. They can receive voice guidance to a parking space and pay through a mobile phone. City planners can gather data on parking occupancy, revenue, and enforcement, and these reports can assist with pricing decisions.[65]

The city benefits with increased revenue through higher enforcement-officer productivity, improved capture rates, and tickets issued for parking violations. Drivers benefit by reducing the time they spend searching for a parking spot, saving them time and money. Everyone benefits through less overall congestion and general improvement in the quality of life.

TRANSPORTATION EXAMPLE: INTELLIGENT TRANSPORTATION SYSTEM, POZNAN, POLAND

The City of Poznań, Poland, implemented a new Intelligent Transport System (ITS) to optimize the overall traffic conditions in the city. A system collects and sends real-time traffic data from more than 200 measurement points (e.g., intersections, parking lots, public transportation) to the traffic management center, where they are evaluated and processed. The ITS provides information on the current traffic situation, available parking spaces, and the departure times of nearby public transportation. Where needed, traffic light control (involving 100 traffic lights) influences the speed of the flow of public transport. More than 85 information panels at the bus and tram stations indicate current information on the arrival time of connected vehicles. All traffic information is also forwarded to the Internet and enabled for radio stations, texting, and e-mail services. At strategic locations along the roads, 12 variable message signs provide essential information for drivers.

This project has increased transport efficiency in Poznań. By enabling better and more efficient public transport, it has reduced the number of individual car users, thereby lowering traffic congestion. It provides pertinent information to users of public transportation, as well as to drivers, thus promoting traffic safety. Also, it contributes to environmental sustainability by reducing fuel consumption and the emission of exhaust fumes from automobiles.

There were specific challenges in the course of implementing this project. First, there was a shortage of qualified and experienced ICT staff in the municipal office, as well as in municipal transport entities that collaborated with the municipal government. They overcome the problem by cooperating with scientific and research institutions, as well as getting technical support from private sector ICT entities.

Second, there was the issue of interoperability between new systems and legacy systems to obtain synergy. The lack of interoperability turned out to be the most critical issue, since the implementation of other projects occurred at the same time. It was necessary to avoid the use of different systems and on-board computers in various projects. Also, to mitigate this risk in the future, it was essential to ensure the openness of the communications protocols.

6.6   Promoting Safety and Security

An important area where Smart City solutions can enhance municipal operations is in promoting safety and security.

One of the significant responsibilities of a city is to protect the public. The public safety system encompasses police, fire department, and emergency ambulance services, disaster prevention/response agencies, the courts, and prisons. It may also include infrastructure and resources from other city departments and non-city agencies. Smart City ICT technologies can contribute to enhancing the safety and security of a city.

Mainly, by maintaining public safety calls for the use of intelligence— information to help people in making more educated decisions--ICT- enabled Smart Cities can help improve public safety outcomes in several ways:

- Generating public safety data
- Accessing the universe of relevant data
- Correlating data sources to create intelligence
- Delivering intelligence to decision-makers

A database may include records such as fingerprints, mug shots, video surveillance footage, and others. However, masses of data alone are not sufficient to create actionable intelligence. Raw data needs to be interpreted and correlated.

Further, one department may require the data from another department, so different departments need to cooperate and share data. For example, police might need access to drivers' license records from the licensing department. Upon creating the information, people who need it must be able to access it easily. ICT, data management systems, and interoperability can help achieve this goal. Smart City technologies, including computing power and analytics, can correlate data, create intelligence, and deliver it to decision-makers.

Some improved public safety outcomes resulting from Smart City technologies can include:

- Reducing response times. Public safety personnel can obtain an immediate and real-time understanding of incidents so that they can respond more quickly.
- Lowering crime rates through more prevention. Analytics can enable the identification and prevention of threats before they cause harm.
- Mitigating pain and suffering.– Here we have an intangible benefit, but it offers real value in a community.
- Enabling operational savings and better deployment of resources. In one study, a hypothetical public safety agency could save 17% of total operating costs through smart technologies and best practices.
- Avoiding criminal justice and correction costs through crime prevention. By applying analytics to an integrated public safety information database, a city can increase its crime prevention capabilities.[66]

Table 5 summarizes Smart City technology enablers and the elements that it can improve.

Table 5: Technology Enablers and their Deployment and Use

Technology Enabler	How to Deploy and Use ICT to Enhance Public Safety
Instrumentation and control	• Implement optimal instrumentation
Connectivity	• Connect devices with citywide multi-service communications
Interoperability	• Adhere to open standards • Use open integration architectures and loosely coupled interfaces • Prioritize the use of legacy investments
Security and privacy	• Publish privacy rules • Create a security framework • Implement cybersecurity
Data management	• Create citywide data management, transparency and sharing policy
Computing resources	• Consider a cloud computing framework • Use an open innovation platform • Have access to a central GIS • Have access to comprehensive device management
Analytics	• Achieve operational optimization • Achieve asset optimization • Pursue predictive analytics

Source: Smart Cities Council, Smart Cities Readiness Guide, * Public Safety.

An example of a city employing a Smart solution for safety and security is Seville, Spain, and in Chapter 9.

6.7   Human Capacity Building

This section addresses the impact of Smart City solutions in terms of human capacity building, covering not only jobs created, but training and skill development.

Building human capacity is an essential element in successfully implementing a Smart City program. Adequate human resource skills must be available. As mentioned previously, it is people who design, implement, operate, and evaluate ICT systems. Managers set objectives, schedules and milestones, manage and oversee the implementation of a project, monitor execution, and evaluate outcomes. Leaders set overall goals, motivate others, and mobilize other resources.[67] Of course, ICT skills are necessary for the operation of conventional city functions. However, the Smart City element adds another layer.

In general, a significant portion of jobs involved in creating and maintaining smart infrastructure requires a good foundation in STEM (science, technology, engineering, mathematics) education. An essential characteristic of a Smart City is that it calls for multi-disciplinary teams working together to plan, design, and operate Smart City systems.

6.8 Education

Skills and intellectual capital are becoming more and more critical in the modern economy, and schools play an essential part in developing these skills. Thus, education is a vital element in economic and social development. The ability of an economy to grow and innovate requires quality education for everyone. Cities need a skilled workforce with relevant training and experience to continue their development growth. Cities are centers of learning, and the Digital Age offers new ways of learning. As suggested by Williamson,[68] Smart City technologies applied to education can in turn help educate and train people to be able to use and work with the Big Data that are produced by Smart City projects so that, as citizens, they will be able to participate in the Smart City themselves. For example, commercial vendors (e.g., IBM and Microsoft) are developing a vision of education as a highly coded, software-mediated, and data-driven social institution. Students can be educated to become "computational operatives" and learn to code to become Smart Citizens in the governance of the Smart City.

> **Smart schools are imagined educational institutions that will contribute to urban governance by shaping citizens' capacities to contribute to the management and optimization of the future of the city.**
>
> - Ben Williamson
> "Educating the smart city: Schooling smart citizens through computational urbanism
> *Big Data & Society*
> December 2015

Commercial companies also see a business potential for big data technologies and practices in education, and they have responded with new developments. For example, Pearson, a vendor of educational products, has established a Center for Digital Data, Analytics, and Adaptive Learning. This facility envisions education systems where teaching and learning become digital, and data is available not just from annual tests, but also from the daily activities of individual students.

Schöenberger and Cukier[69] envision that Big Data will reshape learning through "datafying the learning process" in three crucial ways:

- Through real-time feedback on online courses and textbooks that can "learn" from how they are used and "talk back" to the teacher.
- Individualization and personalization of the educational experience through adaptive learning systems that enable materials to be tailored to each student's individual needs through automated real-time analysis.
- Probabilistic predictions generated through data analytics that can harvest data from students' actions, learn from them, and make forecasts of individual students' probable future performance.

These researchers imagine school as a "data platform" where the real-time "dataficacion" of the individual is becoming the "cornerstone of a big-data ecosystem" and in which "educational materials will be algorithmically customized" and continuously improved."

Companies like IBM and Microsoft are developing Smart Education and Smart Learning programs, and various organizations are beginning to produce materials that envision education as a smart social institution situated in new, digitally mediated urban infrastructure. For example:

- IBM has established a Smarter Education program as part of its global Smart Cities agenda.
- Microsoft's CityNext program features an Educated Cities program, assuming an interdependence between the future city and the school.

The issue is not merely reshaping the school to enable it to track and monitor students but is about educating the Smart City itself by allowing the students to become active citizens who can participate in the practices and performance of planning, implementing, and operating a Smart City.

Many Smart City programs emphasize the idea of Smart Citizens, reasoning that the economic, cultural and political functioning of Smart Cities will rely on smart people who can contribute to the monitoring and management of the city itself. For example, the Milton Keynes smart city program (in the United Kingdom) was a collaboration between the local government and the Open University, called MK:Smart.[70] The program included a major educational initiative known as the Urban Data School. The objectives were:

- To teach young people data literacy so that they can access and analyze urban data sets.
- To create tools and resources to bring data skill education into the classroom.
- To encourage new forms of "active citizenship" through using data to design and evaluate Urban Innovation Projects.
- To devise effective solutions on the local, urban, and global level.[71]

The MK:Smart project finished in 2017; however, a follow-up initiative is in the works. Run by computational urbanists with a focus on educating the Internet-of-Things generation; the program seeks to support learners to acquire the awareness and skills relevant to smart cities.

6.9   Health Services

Smart Healthcare refers to integrating intelligent and networked technologies into health services. Such systems can enable a broader view of overall care, wellness management, and healthy living by monitoring the health conditions of citizens and shift the focus from cure to prevention. It can help ensure the availability of appropriate health services and resources at the right time, and it applies to both in-patient and out-patient environments.

Smart City technologies applied to health services can enable remote diagnosis, remote treatment, digital health records, home health services, and remote patient monitoring. For example,

- Sensors, devices, and smartphone apps can help enable patients to collect data that can be used to monitor and support therapies.
- Data on patient vitals can be collected remotely for diagnostic purposes.
- Mobile platforms can display/transmit electrocardiogram signals.
- The Smartphone can be converted into a patient-specific device to measure, display, and transmit the data generated from sensors.
- Sensors can detect, measure, display, and transmit blood glucose levels.
- Patients can be automatically alerted to take their medication or make an appointment for a health checkup.

EXAMPLE: SMART HEALTH SYSTEMS IN SINGAPORE

In Singapore, a pilot is under way for a Smart Health-Assist program. Sensors are used to monitor the patient's living environment, transmitting the collected data to medical professionals. These sensors can monitor vitals such as blood pressure, heart rate, blood oxygen, and blood glucose levels, among other things. Intelligent sensors in pillboxes, or even in the pills themselves, can remind patients to take their medication at the specified time. All the data is collected and transmitted to the physicians or healthcare providers in real-time, which provides timely information on the patient's progress and state of health.

The technology is provided by Medic-Mobile, which offers smartphones and other mobile devices that operate on the Internet to help health workers extend their reach in the areas of tracking the spread of disease, immunizations, pregnancies, and inventory/stock of medical supplies and drugs.[72]

Smart Health technologies benefit both consumers and healthcare providers. In addition to being able to provide services more efficiently and at a lower cost, the healthcare system can reach patients who previously may not have access to care.

The above case is an excellent example of solutions that are implemented by the private sector without government participation.

6.10 Inclusivity

Governments are instituting programs to enable more people to access government information and services, particularly vulnerable and disadvantaged groups. In its 2016 e-Government Survey, the UN mentions that many countries have established open government catalogs, to simplify the finding of relevant information and to facilitate the ease of use. Some also provide tools such as online guidance and tutorials and notices about the release of information to help people become aware of what information might be useful to them. Others offer workshops and training courses to increase the capacity of people to access information and make effective use of it.

6.11 Sustainability

The UN has adopted the 2030 Agenda for Sustainable Development, among other commitments, to create peaceful, just, and inclusive societies. It points out, however, that achieving this goal will be difficult without adequate institutions. Moreover, ensuring effective, accountable, and inclusive institutions is one of the goals of 17 Sustainable Development Goals (SDGs). Governments are increasingly implementing e-Government solutions as tools to help them "promote policy integration, enhance public accountability, promote participation for more inclusive societies, as well as ensure equitable and effective public services for all..."[73]

Different policy domains may be closely interlinked. For example, enhancing food security involves environmental sustainability, as well as improving rural livelihoods. As pointed out by the UN in its 2016 e-Government survey, "Sustainable development challenges require a holistic and integrative response."[74]

E-government technologies enable governments to integrate different policy domains and their impact on each other, such as food security, sustainable agriculture, and biodiversity. The use of Big Data analytics helps to provide an understanding of the interactions of different policies to formulate strategies based on an understanding of the range of resulting impacts. By merging data from siloed domains, it is possible to obtain a more illustrative picture of the challenges and opportunities for sustainable development.

For example, Rio de Janeiro has a Smart City program for integrated service delivery across the dimensions of sustainable development. The system collects analyzes, and channels data to make better decisions (frequently in real-time) at the municipal level through improved use of technology.

7 EVALUATING SMART CITY PROJECTS

As it happens for any project, we need to assess the outcome of Smart City initiatives, to determine if it was worthwhile. Evaluation offers lessons that can be learned and applied to future efforts. Did a project succeed? If a project failed, why did it fail? Can the problem be remedied to improve future projects? Alternatively, should that type of project be abandoned altogether?

Before saying whether a project succeeds or fails, there must be some expectation of what the outcome will be, and metrics must be established to measure the results and determine the value of the project.

As Cresswell and Sayogo[75] point out, a fundamental question arises about public value assessment: What constitutes good evidence of public value impacts? The outcome of a project may yield value of various types, including financial, political, social, strategic, ideological, quality of life, and stewardship. The different values may produce economic or non-economic outcomes, and each type of value can encompass multiple dimensions or variables. Further, the impacts of a project are not merely isolated events; instead, they are embedded in a context of social and economic activity and may have secondary effects. There is also a difference between projects that deliver direct quantitative results in contrast to those that create capabilities that have the potential to provide multiple streams of qualitative benefits

Some metrics are straightforward and produce results that can be measured quantitatively, such as monetary savings by an agency when it implements a project, or an increase in Gross Domestic Product to measure the economic impact of a project. Other outcomes, such as social and quality of life impacts, are evaluated through non-economic metrics.

For example, the U.S. Government has established guidelines for evaluating its IT investment decision-making for federal agencies. The guide notes that IT investments can have a dramatic impact on an organization's performance. Carefully selected, focused, and well-managed IT investments can dramatically improve performance while reducing costs. On the other hand, poor investment, inadequately justified or poorly managed costs, risks, and benefits, can hinder and even restrict an organization's performance.

Too often, government IT projects have cost too much, produced too little, and failed to improve mission performance significantly. At the same time, it is also generally agreed that the government's ability to improve its service and performance depends heavily on how well it integrates IT into the government's fundamental business/mission. It is a crucial goal of the Clinger-Cohen Act that "agencies should have processes and information in place to help ensure that IT projects are being implemented at acceptable costs, within reasonable and expected time frames, and are contributing to tangible, observable improvements in mission performance." [76]

The General Accounting Office's (GAO) Guide for Evaluating Federal Agencies' IT investment decision-making provides a structure for evaluating and assessing how well an agency is selecting and managing its IT resources. Assessment is approached on three levels: processes, data, and decisions:

- The methods that the organization is using to select, manage, and evaluate its IT investments.
- The data (cost, benefit, and risk) that are being used to make IT decisions.
- The IT decisions that are being made using defined processes and data.

The Guide sees the IT investment management process as an integrated approach with a structured process, providing a systematic method for agencies to minimize risks while maximizing the return of IT investments. The three phases are Select, Control, and Evaluate. Each stage is not a separate step, but each is conducted as part of a continual, interdependent management effort, as illustrated in Figure 24. Information from one phase is used to support activities in the other two stages

Figure 24: Phases in U.S. Government IT Investment Management Process

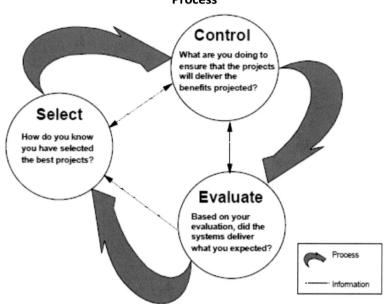

Source: United States General Accounting Office, Accounting and Information Management Division, Assessing Risks and Returns: A Guide for Evaluating Federal Agencies' IT Investment Decision-making, Washington, D.C., Version 1, February 1997, p.6.

Once selected, all projects in a portfolio are consistently controlled and managed through progress reviews. The Control phase helps ensure that, during the course of implementation, the project continues to meet mission needs. If it does not, for example, if investment costs rise or if problems come up, steps can be taken to mitigate the effects of changes in costs or risks. Finally, once fully implemented, we need to evaluate actual versus expected results from a project.

7.1 Economic Evaluation Metrics

City planning frequently involves ICT projects and requires justification. Cities do not merely spend money on technology; they expect a return on their investment—not necessarily as profits, but in terms of cost savings or social benefits, to name a few.

However, how to evaluate the impact of technologies applied to Smart City? Gross Domestic Product and Return on Investment are the usual methods of economic evaluation. The analysis quantifies the savings to individual government agencies and the impact on the government as a whole, as well as the overall impact on the economy.

For example, ASTRO Systems carried out a project to develop e-Government infrastructure in Costa Rica for the Technical Secretariat for e-Government. Although it was a national government project, the development of scenarios for evaluation is the same for municipal projects. Among the tasks, the analysis addressed aspects related to the economic and financial sides of the proposed connectivity infrastructure. The infrastructure consisted of a nationwide broadband network and a shared data center that would serve the participating government entities. Ten government entities participated in the initial selection.[77]

The study evaluated two principal scenarios, as well as a third scenario considered less likely. Option A involved the construction of a new backbone network and data center. Option B analyzed the utilization of excess capacity on existing networks and the development of a new data center. In the third scenario, the Secretariat, acting as a wholesaler, would negotiate and obtain bandwidth and provide it to participating institutions.

The report included forecasts for capital and recurring costs for the three scenarios, along with revenue estimates. It also provided cash flow analysis and Internal Rate of Return (IRR) calculations for the various scenarios with varying discount rates. In addition to being financially viable under most situations, it concluded that the project would have other economic benefits.

Concerning economic efficiency, the study found that the economic benefits of the connectivity infrastructure project would be significant for the government as a whole. In particular, one significant benefit would result from the consolidation and rationalization of the connectivity infrastructure, with the corresponding elimination of redundancies and inefficiencies inherent in the existing situation. There was an expectation that replacing multiple individual networks in use by the different public sector organization, would bring a better allocation of costs for connectivity services and value-added inherent in the data center. Efficiencies would generate additional savings because it can:

- Increase time savings for the public institutions in e-Government, due to new services enabled and the consolidation of the services or various small data centers, many of them in unsatisfactory condition.
- Improve the quality of service, reliability, security, and recovery in emergencies or catastrophic events.
- Integrate public sector systems and services.
- Make available adequate and secure space for public sector ICT equipment and systems.
- Provide monitoring and proactive intervention capabilities.
- Guarantee uninterrupted service.
- Have available sufficient specialized personnel.

These various sources of efficiency create an improved, more efficient economy at the national level as the consequence of better, more transparent and faster procedures for the citizenry and businesses in the various realms of interactions with Costa Rican public sector institutions. Estimates show that a country that implements integrated e-Government initiatives can see economic benefits on the order of 0.5% of the GDP. Under that assumption, the economic benefit to Costa could rise to over US$200 million. Since the Smart City concept is closely related to e-Government, we can expect a similar significant return for projects involving a local area.

The savings to individual citizens present another metric for evaluating the impact. Traditionally, when applying for a permit, license, or any government-required document, the citizen has to visit the appropriate government office, and stand in line, perhaps for an hour or more. Assuming that the citizen works 2,000 hours per year (50 weeks per year, five days per week, 8 hours per day), one hour of waiting represents 0.05% of the citizen's annual working time. If we multiply the average number of hours that each citizen has to wait in line during a year for government services by the total number of people who had to do the same and then multiply the result by the average hourly salary, we obtain the total loss. That loss can amount to billions of dollars at a national level.[78]

How to measure this loss of GDP and subsequent savings produced by ICT-enabled systems? For example, in Colombia, the procedure for obtaining a "Certificado Judicial" changed from the office visit/stand in line model to an on-line process. The savings per individual seeking such a document was 2 hours 10 minutes. From August 31, 2009, to November 7, 2008, there were 371,079 requests for such certificates. The estimated savings to individual users amounted to US$4.54 in indirect costs, giving a total saving of almost US$1.7 million.[79]

7.2   Non-Economic Metrics

In addition to economic evaluation of Smart City initiatives, we can use non-economic metrics, such as the ones proposed in various ISO standards and by the Social Progress Index.

7.2.1   ISO

The International Standards Organization (ISO) is an independent, nongovernmental organization that develops and publishes international standards. Membership is composed of the national standards bodies of 162 countries. The members share knowledge and expertise to develop voluntary, consensus-based, market-relevant international standards to support innovation and to offer solutions to world challenges.

The ISO works closely with the International Electrotechnical Commission (IEC) and the International Telecommunication Union (ITU). In 2001, the three organizations formed the World Standards Cooperation (WSC) to strengthen their standards systems. ISO also has a close relationship with the World Trade Organization (WTO) and works with United Nations partners. In all, ISO collaborates with over 700 international, regional, and national organizations in the standards development process, sharing expertise, and best practices.

ISO 18091, published in 2014, provides local governments with guidelines for achieving reliable results through the application of ISO 9001:2008. This standard sets forth the requirements for a quality management system when an organization:

- Needs to demonstrate its ability to consistently provide products and services that meet customer and applicable statutory and regulatory requirements; and
- Aims to enhance customer satisfaction through the effective application of the system.[80]

For a government to be effective, the citizens need to have confidence that the government and its institutions represent them and work in their interests. To retain citizens' trust, the government must maintain high quality in the services it delivers to the community. ISO 18091 provides a diagnostic model that a local government can use as a starting point and then sets forth guidelines for the implementation of an integrated quality management system that will achieve reliable local government.[81]

ISO 37120, issued in 2014, is related to the sustainable development of communities; it defines and establishes methodologies for a set of indicators to guide and measure the performance of city services and quality of life. It can work in conjunction with ISO 37101. It is part of a new series of international standards in the process of being developed to achieve a holistic and integrated approach to sustainable development and resilience. It offers standardized indicators that establish a uniform approach to what is measured and how to take measurements. The indicators can track and monitor progress on city performance. However, the list does not imply any value judgment or target value for the indicators.[82]

An even newer standard is ISO 37122, *Sustainable cities and communities-Indicators for smart cities*. Just Issued in July 2019, it specifies and establishes definitions and methodologies for a set of indicators for Smart Cities. In conjunction with ISO 37120, it is designed to provide a set of indicators to measure progress towards a Smart City.[83] Later in 2019, ISO 37123 is expected to be published: *Sustainable cities and communities-Indicators for resilient cities*.

7.2.2   Social Progress Index

The Social Progress Index provides a framework to measure various parameters of social progress so that countries can benchmark their success and seek to advance to greater wellbeing for their people. The Social Progress Imperative developed this Index. They are an organization that aims to make the concept of social progress as relevant as the GDP.

Traditionally, the primary measure of a nation's success has been the GDP. The Social Progress Index, on the other hand, considers multiple dimensions of wellbeing: It examines social and environmental indicators that encompass three primary aspects of social progress: basic human needs, indicators of wellbeing, and opportunity. The category for basic human needs covers nutrition and primary medical care, water and sanitation, shelter, and personal safety. The indicators of wellbeing include access to basic knowledge, access to information and communications, health and wellness, and ecosystem sustainability. The opportunity category encompasses personal rights, personal freedom and choice, tolerance and inclusion, and access to advanced education.[84]

The index is designed to reveal the social and environmental aspects of the society rather than purely monetary measures. For example, the indicators measure things like women in school rather than family income.

Further, it measures outcomes rather than inputs, assessing performance in indicators such as access to electricity rather than levels of funding. The design also seeks to be relevant to all countries, measuring the performance of societies at all income levels everywhere. Of most importance, the index offers a practical tool; leaders can act on the information that the index provides. For example, if the index finds that access to electricity in a given country is at a low level, programs can be mounted to increase access to a target level.

The Social Progress Index is being applied to numerous situations to solve problems. For instance, in Paraguay, the national government is setting targets for Social Progress Index performance in addition to GDP targets.[85]

A study in Colombia[86] measures the social progress rankings of 10 Colombian cities in 2014, as indicated in Table 6.

Table 6: SPI Ranking for 10 Colombian Cities in 2014

Cities	SPI for Cities 2014	Social Progress Level
1 Manizales	75,52	High
2 Bucaramanga	72,95	Upper Middle
3 Medellín	72,58	Upper Middle
4 Bogotá	70,33	Upper Middle
5 Pereira	66,63	Upper Middle
6 Ibagué	61,64	Lower Middle
7 Barranquilla	60,07	Lower Middle
8 Cali	54,27	Low
9 Cartagena	53,61	Low
10 Valledupar	52,17	Low

Source: #Progreso Social Colombia, *Social Progress Index for Cities of Colombia*

The study notes that income alone is not necessarily associated with other aspects of wellbeing. While Bogotá is the city with the highest GDP per capita in Colombia, it is only in fourth place in SPI ranking. While it leads in access to primary medical care and education, it exhibits shallow social progress scores in metrics related to the disabled, displaced, and demobilized. The city needs to work on aspects such as opportunity and civil rights, among others.

Manizales, on the other hand, has only two-thirds the average income as Bogotá, but it is most efficient in transforming economic resources into overall wellbeing for the population. This city is particularly strong in the area of Opportunity, and it also scores very high in Shelter and Public Services, Water and Sanitation.

The Social Progress Index for Cities can generate actionable diagnoses for each of the Colombian cities based on a holistic, dynamic, and contextual point of view for twelve areas of human wellbeing: from education to basic and advanced healthcare, to personal rights or individual freedom; from water and shelter to access to information and environmental sustainability; from personal safety to coexistence.

- Social Progress Index for Cities of Colombia
#Progreso Social Colombia

8 BUSINESS MODELS

There are various business models to establish Smart City projects. In the United States, traditional municipal functions such as fire and police departments are generally owned and operated by the city. In many other countries, fire and police may be a provincial, state, or even a national organization; but in the end, they serve the city. Other services such as bus systems and garbage collection may be implemented and operated by the municipality or may be operated privately under a concession. For example, in Dallas, Texas, residential garbage collection is run by the local town, while in many other cities, private companies perform this service under contract.

As seen in Table 7, the models include public ownership and operation, public-private partnerships (PPP), or private ownership and operation. Smart City programs can operate under any one of these models, depending on the nature of the initiative or project. In addition to Smart City projects implemented by city governments with public funding, Smart City projects may be PPP or private enterprise. Even projects entirely in the private sector can help make cities smarter by making functions such as transportation and waste collection more efficient, reducing the use of scarce resources, and helping to preserve the environment.

Further, the business model for a given function or program can be different in different cities. For example, in Dallas, Texas, the Dallas Area Rapid Transit (DART) system is a public operation. On the other hand, the busses in Santiago, Chile, were for many years owned and operated by private enterprise in a competitive environment. Now the city is divided into several areas, and one company has a concession for each area.

Table 7: Business Models for Smart City Programs and Projects

Business Model	Type	Characteristics	Examples
Public	Monopoly	Owned and operated by government	• Traffic Management • Fire Department
PPP	Regulated	Ownership and operation shared between public and private sector	• Infrastructure development/ improvement • Bike Share systems in certain cities
Private	License	Private ownership and operation under license from government; multiple licensees compete	• Taxis • Street vendors
Private	Concession	Private ownership and operation under exclusive concession with no competition	• City garbage collection • Bus routes
Private	Unregulated or minimum regulation	Any private individual or company can offer product or service	• Bike Share systems in certain cities • Apps with city information for Smart Phones
ESCO	Private operation	ESCO works under a performance-based contract with the government authority	• Streetlights

Source: Authors

Public programs implemented by the local government generally apply to functions that the city already operates such as fire department, police, street lighting, and drinking water systems, among others. Smart City projects can also be under public ownership and operation. For example, in 2013, the City of Boston launched a project designed to capture and analyze data from the city's 350 facilities, 850 traffic signals, 64,000 streetlights, and 3,100 city vehicles. The motivation for this project was to reduce the city's US$55 million annual electric bills.[87]

Private ownership can take three primary forms: private ownership/operation under a license from the government, private ownership/operation under a concession from the city, and free enterprise where any individual or company can offer a product or service with little or no government regulation.

With private ownership/operation under a license, there are generally multiple licensees, and they compete. An example of such a model is the taxi services. Most cities have numerous taxi companies, and they compete for business. In Dallas, Texas, some 16 taxi companies offer service.

With private ownership/operation under a concession, generally the authorities grant an exclusive permit to operate in a specific region, and there is no competition for the same area. A typical case is garbage collection from households. The garbage collection company gets the exclusive right to collect garbage. (A city may grant more than one concession, but each concession has its exclusive territory.)

Free enterprise private ownership and operation means that any individual or company can enter a market and compete. For example, the rideshare market has multiple companies, including Uber and Lyft, among others. They have been operating without government regulation, and they have become popular. However, issues have arisen, and various cities are seeking to develop a regulatory framework for this service.

Of particular interest are the public-private partnership and the ESCO.

8.1.1   Public-Private Partnership

In the private ownership model, a private entity owns, operates, and maintains the system, obtains funding in the financial markets, and runs the business to maximize profit for the benefit of the shareholders. The outcome may not be in accord with political, socio-economic goals. For example, private companies would not have an incentive to provide service in poor areas if it would not be profitable. Private companies left to their own devices may not have the resources or the inclination to commit to a project where there is no certainty that it would be profitable, or where the profits are too far in the future. Further, private companies are driven more by short-term profit and loss considerations.

This is where the public-private partnership comes into the picture--To balance the need for profit and the obligation to serve the city needs.

In the scenario of a public-private partnership, a government-owned entity is the owner and operator of the project, but some portion of the ownership or operation is turned over to a private company. Public-Private Partnerships can be a solution where private companies have no interest or operations are not yet financially viable for some reason. Under this model, the government commits resources and capabilities, particularly financial resources. Government investment can make possible a project that private enterprise would not otherwise undertake. The public sector supports the economic needs of the project, while the private sector generally handles operation.

8.1.2   ESCO

An ESCO (Energy Service Company) is a private entity that works under a performance-based contract with the government authority, such as a municipality. An ESCO offers energy service to clients for a fee, generally less than what the client (city) currently pays. The ESCO usually implements comprehensive energy efficiency retrofit projects to reduce costs, and the ESCO retains the savings. In essence, the ESCO acts as a project developer; it handles the project design, financing, installation, and operation. Contracts typically range from seven to 20 years, depending on the implementation requirements. The unique element in an ESCO contract is the guarantee of energy savings.

When a municipality works with an ESCO to implement a street lighting project with new technology, it is the ESCO that invests, and this model is in regular use in the European Union. For example, Wellness TechGroup, a Smart Cities technology company in Spain, works with ESCOs in several municipalities in Spain.

Because of the importance of ESCOs in the implementation of energy efficiency projects, the EU developed a code of conduct for them. This effort supports the development of a trustworthy market in Europe. Rules are tailored to the specific conditions of individual countries and help to increase the transparency of the market and maintain a high quality of energy services provided.

9   CASE STUDIES

In previous chapters, examples demonstrate how targeted Smart City projects are benefitting specific services or sectors. In this chapter, we provide case studies that show the overall strategy and program for selected Smart Cities.

Just as there is no single definition for a Smart City, ranking cities as to the level of their achievement as Smart Cities is equally tricky. Several organizations have formulated a list of the top Smart Cities in the World. For this book, we look at the United Nations' Local Online Service Index (LOSI). [88]

LOSI analyses the municipal Web portals of 40 cities based on 60 indicators related to technology, content, services, and participation and engagement. The results enable a comparison to assist municipal authorities to assess their Web strategy, improve their operations with best practices, and guide policymaking.

The 2018 LOSI Survey was an outgrowth of the UN's e-Government Development Index (EDGI), which ranks 193 countries on their e-Government efforts.

Table 8 shows the top 15 ranked LOSI cities.

A few of these cities will serve as case studies. They are among the top 15 LOSI cities, except for Seville, which is as an excellent example of a Smart City at Level 3.

Table 8: Top 15 LOSI Cities

Rank	UN LOSI 2018	Total Indicators
1	Moscow	55
2	Cape Town	53
2	Tallinn	53
4	London	51
4	Paris	51
6	Sydney	50
7	Amsterdam	49
7	Seoul	49
9	Rome	48
9	Warsaw	48
11	Helsinki	47
11	Istanbul	47
11	Shanghai	47
14	Madrid	46
14	New York City	46

Source: Authors, based on data from *UN E-Government Survey 2018*.

9.1 Cape Town: Smartest City in Africa

Cape Town, South Africa, in second place in the LOSI ranking, is known as the "Smartest City in Africa."[89] The city's vision[90] is "to build a City for all, a City in which no-one is left out." Cape Town has several Smart City activities, including an open data portal, public Wi-Fi, a Smart Grid, and hundreds of cameras throughout the city for public safety applications.

- E-government, which offers more efficient services and better access to them
- Developing ICT skills to promote socio-economic development
- Improved broadband structure and public Wi-Fi to reduce the digital divide
- Surveillance cameras for improved public safety
- Open data portal
- Smart grid technologies to support the digital economy

104

The desire for a well-run, safe, caring, and inclusive city that offers opportunity was the base for Cape Town's Smart City program, which is an integral part of overall development plans. The city's Five-Year Plan for 2012-2017[91] involved continued investment in infrastructure to ensure that Cape Town would have the capacity to provide useful and equitable services and be capable of supporting socio-economic development. For example, developing an integrated Transport Management Centre to encompass several activities, such as:

- A Transport Information Centre where the public can obtain information and lodge complaints regarding trains, buses, parking, and other transportation and traffic-related things

- CCTV cameras and variable messaging signs to monitor and inform the public about traffic conditions.

The current five-year plan,[92] running from 2017 to 2022, explicitly sets forth a specific objective to "Leverage Technology for Progress."

This objective aims to use digital technology to transform Cape Town into the most digital city in Africa through investment in digital infrastructure, growing the digital economy, emphasizing digital inclusion and enhancing the city's digital government capabilities, as well as ensure that Cape Town becomes the preferred destination for technology startups in the country.

- City of Cape Town
Five-Year Integrated Development Plan
July 2017-June 2022

The digital initiatives include:

- A broadband project
- A last-mile project
- A digital citizen interaction platform
- An Emergency Policing and Incident Command Platform
- A Big-Data analytical platform.

The implementation of this plan will put Cape Town firmly at the second level and well on the way to the 3.0 level.

The municipally-owned broadband telecommunications network serves the Cape Town metropolitan area. In addition to serving as the city's corporate network, it provides specialized and dedicated VPNs to the needs of particular city departments, and it also provides public Wi-Fi and other public Internet services. This high-speed network will be central to Cape Town's vision to become a "truly digital city."

The last-mile project is increasing access links for connecting buildings such as clinics and libraries. It is also designed to support critical service delivery systems such as water management, traffic lights, police cameras, and smart meters, among others.

The digital citizen interaction platform, among other things, will make City data available to the public through the open data portal and enhance the existing call center by integrating various business information systems

Cape Town also implemented the Emergency Policing and Incident Command (EPIC)[93] a centralized emergency control platform, upon which information is shared, updated, and consolidated between stakeholders. First responders at the scene have access to the latest data via their mobile devices and can thereby take the necessary measures and have better control of the situation.

The Big-Data analytical platform will enable the City to extract useful information from various data sets; this will allow strategic decisions and improve service delivery through the government.

9.2   Sydney: Digital Strategy as Part of Sustainable Sydney

For Sydney, Australia, the Internet of Things is a crucial element of its Smart City strategy. A public network in the central business district employs Low Power Wide Area Network (LoRaWAN) specifications.[94] A gateway installed on top of a significant downtown building serves the area enables various Smart City initiatives.

Sydney has a digital strategy for the city[95] that is part of an overall *Sustainable Sydney 2020* plan for "a green, global and connected city."

Further, Sydney's program is part of a national effort to develop e-Government and Smart Cities in Australia.

> **Continuing to digitally transform the City of Sydney's public services will have a number of positive impacts including cost reduction and service efficiency; and we believe these benefits can only be truly realised if all our communities are digitally enabled, skilled and included.**
>
> - City of Sydney
> Digital Strategy

The vision of Sydney's digital strategy encompasses six priorities:
1. Champion digital inclusion and lifelong learning
2. Create people-centered digital programs and services
3. Digitally transform the city's engagement with its communities
4. Support business to build skills, knowledge, and infrastructure
5. Actively participate in the digital urban renewal of Sydney
6. Be an ethical innovator in the information marketplace

9.2.1   Digital Strategy

Not only is Sydney implementing Smart City technology solutions, but the City is also taking steps to develop its workforce committed to digital transformation, from the executive level on down. Further, the city is working to ensure that citizens and businesses have the knowledge and skills to use the technology and benefit from it.

The basis for Priority 1 is that the future success of the city depends on the people being digitally active. Sydney already offers computer literacy courses and coding classes in libraries and community centers, and Wi-Fi and computers are available for free in libraries. Although non-digital services are still available so that nobody will be left behind, the priority is on digital services and ensuring digital inclusion for everyone. For that reason, the city seeks to ensure that people have the infrastructure they need and the skills to use it by building on existing skills development programs, partnering with formal education institutions, and supporting less formal initiatives to foster digital learning.

For Priority 2, Sydney intends to improve the design, efficiency, effectiveness, and responsiveness of its programs and services. Cities are being asked to do more with less, and the city seeks not only to cut costs but also to meet the changing needs of its communities—not just applying technology to existing services but redesigning them to enhance the experience for users. Citizens and businesses can already perform various transactions online, such as making payments and ordering library books. Improved digital systems will enable individuals to access their data and records and select services to meet their needs and timing. Expectations are that some services will become fully automated and easy to use with simple and intuitive interfaces.

For Priority 3, the city is already using digital platforms (Sydney Your Say) to enhance digital connectivity and engagement and to support the functioning of the democratic process. For example, an online platform is available for people to make comments and give their opinions on government operations. The City seeks ways to improve digital engagement and data sharing by integrating existing platforms with social media and other online facilities.

For Priority 4, among other activities, the city conducts research into the digital competence of small and medium-sized businesses and works with them to enhance their digital skills and capabilities.

For Priority 5, Sydney already provides free or low-cost physical access to digital technology (e.g., Wi-Fi). For the future, the infrastructure is being assessed to determine what extensions and expansions are needed. Also, a plan is underway to implement a network in key areas of the city's public domain for Smart City projects such as data collection through a sensor network.

Among other activities, the city is exploring the potential of street assets (e.g., bus stop shelters) to support the city's digital infrastructure. It is also working on a monitoring and control system for public parks to improve the management of water/energy consumption.

For Priority 6, the City already makes available much of the information it produces and is considering ways to increase the data available. Releasing open data and making it searchable opens opportunities to improve government functions, and it also opens the issue of transparency and privacy. Sydney has an Information and Technology Strategic Plan and is developing a data governance strategy and management framework to set forth the principles and mechanisms for the use of data, including privacy and security.

9.2.2 Australia's Smart Cities Plan

Not only do individual Australian cities have digital plans, but the Australian Government has formulated a Smart Cities Plan.[96] Australia already has some of the most livable cities in the world, and the objective is for them to remain that way. The government believes that collaboration at all levels will work in its favor.

> Smart investment that enables partnerships between government and the private sector will deliver better infrastructure sooner, and within budget constraints.
> However, funding is not enough. The global lesson is that cities collaborate to compete. Success requires all tiers of government, the private sector, and community to work together towards shared goals. Fundamentally, better cities policy starts with a commitment from all levels of government to work together to deliver common goals— including reforms that make our cities easier to invest in and do business.
>
> - - Commonwealth of Australia
> Smart Cities Plan

The Smart Cities Plan gives the Australian government's vision for cities and formulates its plan to maximize their potential through three pillars:

- Smart Investment
- Smart Policy
- Smart Technology

Projects are prioritized to meet broad economic goals and the objectives of the cities themselves, such as jobs, affordable housing, and environmental considerations. Incentives are expected to motivate all levels of government and other stakeholders to invest in critical programs and projects. The intention is to take advantage of new technologies in transportation, communications, and energy efficiency. Further, the plan does not focus simply on the capital cities or the central business district of cities; it covers all cities of all sizes and all areas.

The commitments to back the plan are:

- $50 million for Infrastructure Planning
- Establishment of an Infrastructure Financing Unit
- Inviting State and Territory Governments to partner in programs and projects

The money is expected to accelerate the planning and development of significant infrastructure projects. The financing unit works with the private sector to develop funding and financing solutions. Moreover, partnerships are expected to yield better outcomes through common motives and coordinated actions and investments.

9.3 Istanbul: Advancing to High Prosperity Levels

Istanbul is undertaking a Smarter City initiative. This effort fits into the context of global trends, as well as Istanbul's Regional Plan and Turkey's Tenth Development Plan, and the new 11th Development Plan continues the work.

Turkey has become a middle-income country with a GDP per capita of over US$10,000. Turkey is a member of the Organization for Economic Cooperation and Development and the G20 and is a candidate for membership in the European Union.

Turkey aspires to become a member of the European Union and is working to conform to EU norms and principles. The EU has economic policies, as well as a digital strategy and policies on energy and the environment. For Turkey to gain entrance into the EU, it must comply with these policies.

Therefore, the objectives that Istanbul expects to achieve with its Smarter City Initiative are relevant to advance Turkey's case for entrance into the EU.

9.3.1   In the context of the National Development Plans

The Tenth Development Plan for Turkey, which covered the period 2014-2018, is seen as a milestone in advancing the nation to high prosperity levels, in line with 2023 targets. The goals include high and stable economic growth, as well as issues such as the rule of law, information society, international competitiveness, human development, environmental protection, and sustainable use of resources. The long-term development goal is to improve Turkey's global position and enhance the welfare of citizens with structural transformations. The target for GDP per capita in 2023 is US$25,000, along with raising exports to US$500 billion, reducing the employment rate to 5%, and reducing the inflation rate to single-digit levels. Other expectations include:

This process of transformation is being supported by pursuing developments in the field of science and technology and integrating innovations into the productive structure.

Of interest is the expectation of an increase in productivity in the public sector by e-government applications and services. This area has been a high priority since the framework of Information Society Strategy and Action Plan was promulgated in 2006 during the Ninth Development Plan.

- Numerous e-Government projects were carried out during that period.
- An e-Government Gateway was delivered.
- About 14 million users are registered, and nearly 600 services had been integrated into the Gateway by the end of 2012.
- An Address Registry System and Electronic Public Procurement Platform (EKAP) are now available.
- Pilot schemes of the Central Registry Record System (MERSIS) and Republic of Turkey Identification Card projects are ready, and activities for dissemination launched.

Other application projects cover areas such as justice, education, health, social security, police, and public financial management.

Personal and business use of e-government services has increased, and the level of citizens satisfaction on e-government services reached 94.4% in 2012.

Table 9 shows the developments and objectives for e-Government during the Tenth Development Plan period.

Table 9: Developments and Objectives in e-Government during the Tenth Development Plan

	2006	2012	2013	2018
Individuals Using e-Government Services (%)	26.7[2]	45.1	48.0	65.0
Enterprises Using e-Government Services (%)[3]	66.2	81.5[4]	85.0	90.0
Satisfaction Rate of e-Government Services (%)[5]	95.0[6]	94.4	97.5	98.0
Number of e-Govt. Gateway Registered User (Million Persons)	0.01	13.8	15.0	30.0
Number of Services Provided through e-Government Gateway	22[7]	547	700	3,000

Source: 2006 and 2012 data are from TURKSTAT and TURKSAT. 2013 and 2018 data are estimates of the Tenth Development Plan.
(1) The ratio of persons benefitting from e-government services to internet users of 16-74 age range.
(2) 2007 data
(3) The ratio of enterprises using e-government services to all enterprises having 10 and more employees.
(4) 2011 data
(5) The ratio of satisfied and medium level satisfied persons to total number of e-government services users.
(6) 2009 data
(7) December 2008 data

Source: Republic of Turkey, Ministry of Development, *The Tenth Development Plan 2014-2018*, Ankara, 2014.

Carrying out e-government activities as a part of an integrated public administration approach, along with the need for enhancement of coordination is considered of crucial importance. During the period of the Tenth Development plan through 2018, e-Government the writing of applications and services continued, as well as the integration into the e-Government Gateway. Some outcomes are:

- Interoperability of the information systems of public agencies.

- Improvement of the capacity of public agencies working on e-Government project preparation and management and strengthening of human resources in IT departments.
- Updating of relevant legislation to improve the efficiency of public agencies in ICT procurement.
- New legal, administrative, and technical arrangements regarding the protection of personal data and information security in providing public services.
- Evaluation of products, services, and trends like open source software, big data, cloud computing, green ICT, mobile platforms, and the Internet of Things, with suitable solutions. [97]

Turkey's 11th development plan, just issued in July 2019, is a roadmap to improve the country's position in the international arena and enhance its welfare.[98]

9.3.2 In the Context of the Istanbul Regional Plan

The Istanbul Development Agency (ISTKA) has issued an Istanbul Regional Plan for the period 2014-2023. The overall goal is to transform Istanbul into a strategic actor in the global economy while preserving its unique tenets and improving living standards for citizens: to be a center of innovation, creativity, and high value-added activities, and to expand the transfer of knowledge and technology.

This plan presents an overall development vision for the region for the decade, with strategies and objectives to guide planning and investment decisions for Istanbul. It embodies a broad consensus among public, economic, and social stakeholders on a long-term local development agenda, and the means to achieve objectives. It addresses topics encompassing industrial transformation and entrepreneurship, health and education, institutional capacity and governance, transportation, energy, and environment.

The Plan consists of 23 priority areas, 57 strategies, and 476 objectives and measures to achieve the overall vision. Examples of the strategies used in the Plan are summarized below.

- Strengthening Istanbul's foreign trade
- Creating an industrial structure that uses advanced technologies and employs skilled labor
- Improvement of public transport infrastructure and services
- Enhancing the human resource capacity of Istanbul in the field of R&D and innovation
- Improving accessibility and quality of education
- Provision of accessible, effective, and efficient health services
- Ensuring energy efficiency and clean energy usage
- Ensuring sustainable urban development and smart growth
- Transforming the workforce in line with the changing economic structure
- Providing the sustainable management of basins and water resources
- Improving security and the perception of urban safety in Istanbul
- Enabling the disaster management system to create a safe Istanbul with a high quality of life. [99]

The Smarter City Project is expected to contribute to the attainment of these goals. This Initiative is the responsibility of the Istanbul Metropolitan Municipality (IMM), the public sector entity responsible for general management of the city of Istanbul. Its obligations cover housing, social services, transportation services, health services, environmental management, cultural services, urban and social order management, and disaster management.

The project seeks transformation in seven main domains: environment, mobility, energy, living, governance, economy, and human. To support its efforts, IMM is working to improve and enhance its Smart City capabilities in several areas including disaster/crisis management, citizen 360° services, geographic information systems, geospatial analysis, and a sophisticated browse, search and discovery portal, as well as decision support utilizing Big Data.

The Smarter City Initiative seeks to improve the city's operations through enhanced IT solutions and to develop a cloud-based environment capable of aggregating data from existing municipal databases and information inputs. IMM expects that this Big Data infrastructure will allow it to retrieve data from a variety of domains, which will generate comprehensive analytics to support citizens and improve decision making and planning within the municipality.

To enhance citizen engagement, IMM seeks to develop improved systems for understanding citizen needs and, in turn, develop additional citizen IT services, known as Citizen 360°. IMM is implementing best practices for the provision of citizen information through the establishment of a central data repository to aggregate data from IT domains across municipal entities and services. The objective is to be able to better serve the citizen through the addition of new facilities or the elimination of duplicate or unnecessary services.

IMM uses Geographic Information System (GIS) technology to reduce costs, improve city services, create a better community for citizens of Istanbul, and support disaster and crisis awareness, management, and mitigation. This technology enables citizens to have better information via interactive maps and makes it easier for the Municipality to deliver needed services such as hospitals, clinics, police stations, and other government-based service centers.

IMM provides access to the data stored in a cloud-based environment for a variety of users by offering a portal to disseminate the data. The portal provides a platform for the delivery of expanded or additional services, including a data directory so that citizens can view and access the available municipal data.

IMM is implementing a central repository of Big Data by integrating the data that currently exist in several unconnected systems and databases. The ability to integrate and access the data increases the capacity for city planning and innovation. Once this data is available in a central repository capable of supporting both structured and unstructured data (data from sensors, satellites, social media, mobile communications, e-mail, radio frequency identification, and existing enterprise applications), city planners can utilize the data as a tool, for example, as emergency and disaster planning and decision support.

9.3.3   Implementation

IMM has already implemented many Smart City projects. For example, in 2012, the U.S. Trade and Development Agency (USTDA) funded a contract to provide technical assistance to develop Turkey's Greater Istanbul Municipality for the implementation of a Consolidated Technology Center. The outcome of this effort was the consolidation of over 40 municipal data centers, command and control centers, and call centers operated and maintained by numerous municipal agencies into a joint Technology Center. That effort provided for a comprehensive blueprint for data and systems migration, as well as a plan for the consolidation of resources within the Technology Center.

In 2016, USTDA funded another technical assistance effort[100] designed to expand services in Istanbul, incorporating technologies for transportation, education, health services, power, security, and crisis management. The expansion of capacity supports the Smarter City offerings through the implementation of a Big Data system to support not only the current collection of structured data, but also the collection and analysis of unstructured data such as data from sensors or video. Data from these and other domains need to be integrated and correlated to enable insights to enhance decision making and promote critical actions to make the city smarter. Other projects are also ongoing simultaneously.

Istanbul is also part of the CitySDK (City Service Development Kit) and its application pilots,[101] partially funded by the European Union. The purpose of CitySDK is to foster the creation of applications in smart mobility and tourism, among others.

IMM is working with Ericsson and ISBAK AS (Istanbul Informatics and Smart City Technologies Inc.) to achieve Smart City Goals. IMM established ISBAK AS in 1986 to perform project design and provide implementation services in traffic and systems engineering. Now ISTAK works in over 50 cities in Turkey, as well as more than 20 countries around the world. It focuses on traffic monitoring systems, intelligent transportation systems, and Smart City solutions. Through technology and innovation, IMM and ISBAK seek to attain an integrated transformation in a short time.

9.4 Shanghai: Vision and Strategic Plan to Foster and Guide Development

Shanghai presents a good illustration of the need for a vision and how the strategic plan fosters and guides the development.

In 2013, the Shanghai government issued an Action Plan for building a Smart City.[102] This plan set forth the guiding principle, implementation principles, development objectives, and an outline of tasks. Also, measures were put in place to support the Smart City effort. These included optimizing the market mechanism, fostering scientific and technological innovation, and strengthening organization and leadership, among others.

> **Pressing ahead with building Smart City is for Shanghai an important means of realizing "innovation-driven transformation", a major move to further put into practice "Better City, Better Life", and also the logical call of a new round of fast development of informatization. Since the 1990s, having taken the advance of informatization as a strategic measure governing entire modernization development and made consistent efforts for three consecutive five-year plan periods, Shanghai has now enjoyed a leading position in the country, with some indicators reaching the advanced level of developed countries and a batch of new technological achievements being fully shown or demonstrated at the World Expo, which has laid a foundation for building Smart City. In order to unite efforts to push forward the Smart City drive, this Action Plan is drawn based on the "Outline of the 12th Five-Year Plan for the Economic and Social Development of Shanghai".**
>
> - Action Plan 2011-2013 of Shanghai Municipality for Building Smart City

9.4.1 Action Plan

The guiding principle is to advance the city's goal of being an international economic, financial, trade, and shipping center.

The implementation principles included:

- Proceed step by step from a solid foundation

- Pursue innovative development to benefit people's lives
- Determine priorities and focus on projects
- Municipality and districts coordinate on critical projects
- The government provides guidance, and businesses act as leading players
- Develop objectives
- Outline tasks

The goals involved deploying cutting edge infrastructure to raise the city's capabilities to a world-class level and keep key sectors competitive in the international market including:

- Completion of broadband city and wireless city
- Developing information sensing and intelligent applications
- A new generation of the IT infrastructure to support Smart City building
- Reliable, reliable, and controllable information security

Some of the tasks included WLAN hot spots, an undersea optical cable system, an Internet Data Center, a Super Computing Center, Intelligent Transportation, e-Government, and an Internet of Vehicles, among others.

9.4.2   The Current Platform

Today, Shanghai's Smart City platform integrates several systems, including:

- Community data portal
- Smart City Card
- Community public service integrated reservation system
- Community home care service system
- Community health management service system
- Intelligent property management system
- Citizen's quality practice social certification system
- Learning platform for community innovation education service

Shanghai's Smart City builds on four pillars:

- A database to provide a comprehensive community information base

- The Smart City Card that to be used for various applications
- Integrated management command information platform
- Public service information platform

The database covers the region, people, housing, things, units, buildings, and other dynamic information, providing for browsing, display, statistical analysis, and decision support. The Smart City Card provides identity authentication, consumer payments, and credit to cover the necessities of life through a line of financial services.

The Integrated Management Command Information Platform is designed to improve the regional grid, combined with the IoT, real-time monitoring of community operations and response to various events. The Smart Community Public Service Platform enables residents to make medical appointments and obtain home care, among other things.

Shanghai has substantial Smart City information infrastructure, and the number of applications continues to increase. Fiber-to-the-home coverage already exceeds 95%, and home broadband access bandwidth averages 35.7 Mbps.

Applications encompass traffic, healthcare, and credit, among others. For example, electronics are installed in more than 7,000 bus stations and 500 public parking facilities; several hundred medical institutions are part of the Shanghai Health Information Network; and the Smart Credit information service platform covers 100%. Also, I-Shanghai Public Service (which is operated by a third party) provides information on hundreds of public places such as public transportation and tourist attractions.[103] Companies can purchase open data collected from the city's sensors and cameras, which they then present thru publicly available information tools, dashboards, and other services to enhance the quality of life in the city.

To improve the parking situation, the Chinese tech company Huawei developed a Smart Parking network that enables drivers to find, book, and pay for parking from a smartphone app. Chips are embedded in parking spaces in several hundred parking lots. The system collates and transmits real-time information on parking lot occupancy through the app. The system reduces potential congestion that would be caused by drivers driving around looking for parking spaces.

Over 100 government services for citizens are available through the Citizen Cloud. This platform includes driver license information, healthcare records, and an array of community services.[104]

9.5   Seville: Participating in Spanish Smart Cities Network

Seville is the capital and largest city of the autonomous community of Andalusia, Spain. It is the fourth largest city in Spain. The population of the city is 703,000, and the metropolitan area population is about 1.5 million.

Seville is working on several strategic initiatives involving Smart City solutions. The city is instituting international best practices and has successfully implemented several projects.

Among other activities, Seville is participating in the Spanish Smart Cities Network (RECI). (Spain's EGDI ranking is number 17.) The objective of this network (Red.es) is to exchange experiences and work jointly to develop a model for sustainable management and improve the quality of life for citizens. An action framework promotes the development of pilot projects designed for public services that will mark Seville as a Smart City. It touches on aspects such as energy savings, sustainability, mobility, e-administration, and attention to safety and security.

Seville's Innovation Master Plan[105] sets forth an array of activities designed to foster innovation to develop a model sustainable city based on the intensive use of new technologies. Seville's City Council assumes a leadership role in this effort to improve public services and operations and to make the government more citizen-oriented.

Its Action Plan is a tool to ensure the achievement of the objectives. In general, the plan establishes a strategy to foster sustainability, quality, and efficiency of the municipality of Seville, attracting talent and private investment as economic engines for Seville.

In 2013 Seville entered into a collaboration agreement with the School of Industrial Organization (EOI) to drive innovative initiatives and projects that will promote economic activity and a city model based on knowledge and the use of technology. Of interest is the acceleration of enterprises in strategic sectors in Seville, as well as the development of small and medium-sized companies in Seville's tourism sector. The objective of both projects is to adopt innovative business models and develop viable products as a strategy to accelerate the growth of small and medium-sized companies in Seville and its area of influence.

9.5.1   FI-WARE Project in Seville

One of Seville's significant efforts is an implementation of the EU's FI-WARE Project. Currently, the European Commission is devoting a considerable effort to fostering research into new smart information technologies. The endeavor is an open-source platform for developing the digital future. It is also creating new IoT business opportunities. Industrial production cases, pilot smart city projects, and projects for utilities, are using the technologies. For example, a city water quality pilot in Seville. Expectations for FI-WARE include:

- Drive essential standards for opening up information silos
- Make IoT simpler
- Transform Big data into knowledge
- Tap the potential of right-time Open Data
- Enable the Data Economy
- Ensure sovereignty of data[106]

These items apply to municipal governments as well as national governments

In cooperation with Red.es, Seville is developing FI-WARE. The objective is to put into service and operate a platform with a new generation of Smart City applications, which will enable the city to improve the provision of public services. The system provides an infrastructure with a node in Seville for the development of future Internet services based on the Fi-WARE platform, which expects to have the support of small and medium-sized businesses. The FI-WARE initiative is the central axis for incorporating new elements of innovation in the ecosystem. It is intended to enable improved municipal government, sustainability, and quality of life, economic competitiveness, and innovation.

The FI-WARE Smart City project in Seville is one of the first IPv6 end-to-end (devices, 3G M2M network and IoT platforms in the cloud) trials. It monitors water quality at the city's public fountains. The project's IoT architecture employs multiple open source tools from the FI-WARE platform. The system obtains sensor data from hardware at the fountain source and uses the FI-WARE open-source IoT agent, which feeds the data into a hub, and it can also draw in data from various other sources via APIs. The information then goes through the security API protocols to be available externally and connect the context broker to the dashboards.[107] The architecture is in Figure 25.

Figure 25: FI-WARE Platform

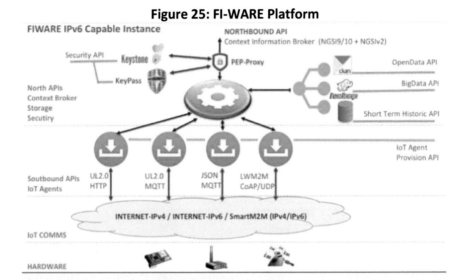

The IoT architecture for Adevice's Seville smart water fountain pilot project. (Image courtesy of Carlos Ralli Ucendo, Telefónica I+D)

9.5.2   LIFE-EWAS Project

EWAS is a European initiative within the LIFE+2013 Environmental Policy and Governance Project Application, which is an international consortium composed of other environmental consulting and waste management companies. The purpose of the project is to test efficient and sustainable methodologies for managing waste, utilizing ICT tools that enable the reduction of greenhouse gases.

As a public sector partner in this effort, the city of Seville, through LIPASAM, is the testbed for a pilot and making use of the results of EWAS. Wellness

Smart Cities & Solutions, a Spanish technology company, provides the technology for this testbed.

LIPASAM is Seville's public sanitary service. It is responsible for 14 sq.km and 1,077 kilometers of roads, as well as the collection and treatment of urban waste. Integral to its work is saving resources and avoiding environmental pollution. To perform its job, LIPASAM employs 1,555 operators, counts 461 vehicles, and manages 50,000 public receptacles.

For this test, they monitored 268 waste containers in Seville, on three collection routes. Sensors were installed on the containers to measure how full they were. Traditionally, collection trucks have defined routes and schedules; they collected and emptied every container on the route, whether it was full or not. This situation offered an opportunity to test ICT solutions that could help this public service achieve improvements.

LIPASAM deployed the Wellness Quamtra system for the test. It consists of hardware devices to measure the level of fullness of the containers and a software platform to visualize and analyze the data collected by the sensors. By using an ultrasonic beam, the sensors measure the level of fullness of the container and collect other useful data. The data are transmitted wirelessly to the management platform and analyzed by the municipal collection manager.

Sensors were installed on the test containers to measure how full they were. The data from the sensors were collected and transmitted to the platform for analysis. With the information obtained through the system, it was possible to reduce the number of times the containers are emptied from three times every twelve days to one time every seven days. The results translate into a reduction from 100 collections annually to 34. It also permits optimization of the routes by changing from three static routes to one dynamic path covering the area. It also reduces the time that the vehicle is on the road by 400 hours annually, for a reduction in noise and odors, as well as improving traffic flow.[108] A scheme of the system is in Figure 26.

Figure 26: Scheme of Quamtra Solution

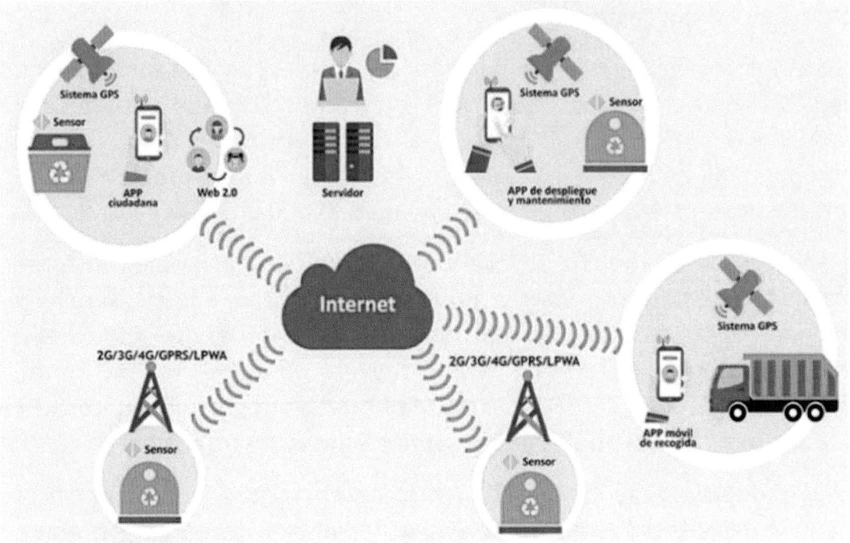

Source: Wellness Telecom Smart City Solutions.

Experience indicates that the system can pay for itself in two years.

9.5.3   Safety in Seville

Seville welcomes nearly one million local and international visitors during the Holy Week religious festival. The ornate processions march through the city at all hours of the day and night, with no shortage of onlookers, crowded among the narrow streets. The Holy Week is an extremely important time of year for the local economy. According to the most recent estimates, the city brings in more than 160 million euros from tourism and increased spending from the local population.

In 2017 the city experienced disturbances during the processions, where mass panic led to several stampedes throughout the city. Nearly 100 people suffer injuries during the incidents, and the city's reputation as a safe destination for tourists and citizens was at risk. To prevent further incidents, the local government initiated an innovative Smart City project for public safety to prevent, detect, and respond to incidents at crowded, large-scale events in real-time.

Looking for a solution to that situation, the municipality of Seville decided to deploy a comprehensive Smart Initiative. The solution integrates multiple systems, using IoT platforms, Smart Street Lighting, PA Systems, and CCTV for people flow analytics. This system enables the coordination of various security forces and the proper use of the city's infrastructure.

The integrated platform enables city managers to have a comprehensive view of multiple city functions on the same management dashboard, thereby allowing faster decision making to act upon the issues that may arise in the City. At this time, some decisions are still made by humans, while others are already autonomous, such as increasing the level of the streetlights when certain conditions are detected. They call that overarching system the "Smart City Brain." Seville is moving into becoming a Smart City with many autonomous decisions (Smart City 3.0).

The fruits from this initiative appeared on the 2018 and 2019 Holy week festivities when no public disturbances or incidents reports in Seville.

A diagram for this type of architecture is in Figure 27.

Figure 27: Architecture Diagram for Public Safety

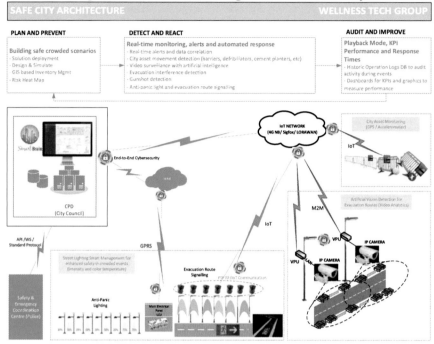

Source: Wellness TechGroup

ABOUT THE AUTHORS

Lorenzo Madrid has more than 30 years of work experience in the public and private sectors, including over ten years at Microsoft Corporation, where he served as its Worldwide Senior Executive in different capacities. As part of his responsibilities, Mr. Madrid oversaw IT project development and implementation for Governments in over 50 countries. His most recent projects for smart solutions in cities include "Istanbul – Smarter City," "Rio de Janeiro – House of Commerce" and "Cape Town – Intelligent AFC System" among others. He also served as Chief Information Officer (CIO) at the Secretary of Education of the State of São Paulo, Brazil, where he was responsible for managing and implementing computerized management systems on behalf of 6 million students and 300,000 teachers. Previously, Mr. Madrid founded Princeton Systems, a Brazilian software development, and distribution firm, serving national, provincial and city government clients through a network of international partners, providing diverse service offerings.

Clients and academic institutions recognize Mr. Madrid for his work in governance, public policy and ICT to promote economic and social development; he was a member of the Advisory Board of the ICEGOV (International Congress of Electronic Governance), organized by the United Nations University. Mr. Madrid has also served as a lecturer at the Lee Kwan Yew School of Public Policies in Singapore, at the School of Government in Dubai, at the Mumbai Management Institute of India, at the Harvard Kennedy School of Business and the FIA-Management Institute at the University of São Paulo. He has presented at numerous Smart City conferences including the Andalucía Digital Week (Spain), Smart City Business (Brazil), the Samos Summit on ICT-enabled e-Governance (Greece), the Future Cities Conference (Slovenia), the Euro-Asia Blockchain Summit (Malaysia), and the Smart Cities Blockchain Conference in Orlando, Florida

Mr. Madrid was appointed as an Industry Fellow by the Centre for Technology in Government at the State University of New York and has published books and scholarly articles on Smart Cities, Government Transformation through IT and about the critical linkage between IT investments and social-economic development. In addition to his current advisory role to Wellness TechGroup, a Spanish company providing advanced solutions for Smart Cities, he is a Program Committee member for the "International Conference on Smart Governance for Sustainable Smart Cities - 2019" jointly organized by the United Nations University and the European Alliance for Innovation.

Linda Lee Bower holds B.A. and M.A. degrees in Economics from The George Washington University. She has been an independent consultant for over 30 years, specializing in infrastructure development and socio-economic development. Her work focuses particularly on telecommunications and information technologies (ICT), energy, water, and transportation. Her work encompasses sector planning and development, policy analysis and regulatory reform, private sector development, market forecasting, and financial projections.

Projects work covers nationwide broadband networks, e-government programs, and Smart Cities, among others. She has done fieldwork in numerous countries in Latin America, Africa, Central Asia, and the Middle East. Some major projects include a Smart City project for Istanbul, a national fiber-optic broadband network in Rwanda, rural telecommunications development in Afghanistan, small hydroelectric projects to deliver electricity to farmers in Uganda, a 20-year telecommunications master plan for Egypt, and airport development in the United States, among others.

In addition to analytical abilities, Miss Bower has provided leadership for multidisciplinary project teams, conceptualizing, and planning complex projects.

She has also published dozens of technology assessment/market forecast reports on various sectors and new technologies as they have developed. These studies assess the technology of interest, analyze the competitive situation, identify applications and user groups, analyze policy and regulatory developments, perform economic analysis, and forecast demand, revenues, and equipment sales, sometimes on a global scale.

She has presented papers at conferences such as the IEEE-ISTAS conference on technology in society and the Samos Summit.

ANNEX—TAXONOMY OF SMART CITY INITIATIVES

Categories and Activities

#	Developments	Cities Examples
1. ICT Infrastructure		
1.1	Implementation of free wi-fi in public areas and municipal buildings	New Taipei City, Taiwan, Brasilia, Heraklion
1.2	Implementation of optical fiber network (MAN)	Midtown Manhattan, New York City, Trikala
1.3	Data centers infrastructure for collecting and storing data from Internet of Things (IoT) sensors	Taiwan, Trikala
1.4	Hardware and software upgrading to the municipal departments for a highly efficient Backoffice	Heraklion
1.5	Electronic document workflow management systems for municipal offices - departments	Athens
1.6	Info-kiosks installation for providing information to citizens and visitors	Tepebaşı, Trikala, Barcelona
1.7	Installation of electronic boards providing information in real time (e.g., weather, local news, etc. on duty pharmacies)	Sydney
1.8	Fiber to the Home network	Singapore
1.9	City Operation and Management Center	Rio de Janeiro, Istanbul

#	Developments	Cities Examples
2. Environment and Agriculture		
2.1	Installation of electromagnetic radiation measurement sensors	New York City
2.2	Installation of noise measurement sensors	Barcelona
2.3	Installation of air pollution (atmospheric microparticles, CO2 emissions) measurement sensors	City of Malaga, City of Amsterdam, Tallinn,
2.4	Installation of rain level measurement sensors	Chicago, City of Amsterdam, Genoa
2.5	Installation of light level measurement sensors	City of Rotterdam, City of Amsterdam
2.6	Seismograph Installation (Installation of Seismographs)	Redwood City
2.7	Installation of sensors for the measurement of Radon	München
2.8	Climate monitoring and forecasting	Genoa
2.9	Garbage Collection Sensors	Seville (Clean City)
2.10	Urban Agriculture Consulting	New York,
3. Transportation - Mobility		
3.1	Actions for traffic management improvement in real time, within/inter municipal areas using specialized applications (Smart traffic Lights)	Thessaloniki, City of Amsterdam, Singapore, La Rinconada

#	Developments	Cities Examples
3.2	Use of intelligent systems at pedestrian crossings for safe movement (Zero Vision Org)	Redmond, WA, USA
3.3	Smart stops (eg online bus arrival marking) for public transportation	Athens
3.4	Installation of sensors/cameras to means of transportation or roads for traffic flow monitoring	Singapore
3.5	Smart information signs for traffic condition	Barcelona
3.6	Car parking spaces sensors providing information and guidance to drivers for parking availability	City of Amsterdam, Singapore
3.7	Application for route tracking of cyclists and reports submission	City of Amsterdam
3.8	Vehicle fleet management	City of Amsterdam
3.9	Automated system for public bicycles	Barcelona, Rhodes
3.10	Collision avoidance sensors	Columbus,
3.11	Low Bridge alarm systems	
3.12	AFC – Automatic Fare Collection & Integrated Ticket Systems (Econ. Development as well)	Visité Paris / Oyster Card (London) / BOM SãoPaulo / Istanbul Card
4. Health		
4.1	Implementation of health care telemonitoring system to support	Singapore

#	Developments	Cities Examples
	vulnerable groups of people (disabled, Alzheimer's disease etc.)	
4.2	Implementation of telemedicine system for measurements of some key indicators (pressure, sugar etc.) on citizens and medical records archive incorporating advice from the hospital / health center doctors	Singapore
4.3	Implementation of applications for remote monitoring patient progress in isolated countryside areas	Trikala
4.4	Establishment of Health Centers	Trikala
5. Waste Management & Water Resources		
5.1	Online quality measurement system of drinking water	Breda
5.2	Online monitoring system, with appropriate sensors detecting possible water and/or sewage leaking	Nottingham
5.3	Online monitoring system for immediate detection of possible water leaks in closed irrigation canals or irrigation tanks	Nottingham, Petaluma
5.4	Actions encouraging - informing citizens via tele-education on recycling	Paphos
5.5	Online monitoring and management system of pumping stations	Aveiro
5.6	End to end irrigation management system with dam operation control, pumping stations control, water flow control in tube-like channels	Almare, Petaluma

#	Developments	Cities Examples
5.7	Online waste containers management system (with occupancy sensors) and waste collection fleet management (GPS)	Cambridge
5.8	Decision Support System for sustainable water / waste management.	Bochum, Seville
5.9	Surveillance Application for COC or NOC or Sewage Treatment (Processing) Units	La Rinconada
5.10	Online data collection from hydrometers	Genoa
6. Energy – Sustainable development		
6.1	Photovoltaic installation in municipal buildings and ports	City of Amsterdam, La Rinconada
6.2	Construction of wind farms (plus through crowdfunding)	City of Amsterdam (Ship-to-grid)
6.3	Energy savings in municipal buildings by upgrading exterior wall with insulation claddings and integrated intervention in cooling - heating systems. Energy management system	Singapore
6.4	Energy saving of municipal street lighting and public spaces (e.g. by replacing existent lamps with led type ones, with remote control system). Smart lighting	Barcelona
6.5	Actions on citizen information and awareness via tele-education aiming to energy saving	City of Amsterdam
6.6	Fuel consumption measurement on municipal transportation vehicles aiming to	Heraklion, Trikala

#	Developments	Cities Examples
	fuel consumption reduction by redesigning routes selecting the most appropriate path (fleet management)	
6.7	Smart meters for energy consumption	City of Amsterdam
6.8	Photovoltaics in municipal public areas	City of Amsterdam
6.9	Smart plugs system	La Rinconada
7. Tourism - Culture		
7.1	Design and creation of cultural infrastructure and agents' management system with detailed reporting and promotion via municipal website	Thessaloniki
7.2	Developing electronic local tourist guide	Barcelona
7.3	Developing tourism content applications for mobiles	Barcelona, Trikala
7.4	Protection, promotion and enhancement of museums, galleries, monuments, caves, archaeological and historical sites through virtual tours	Barcelona
7.5	Digitization of museum content for creating digital cultural footprint	Paris, London, Rome
7.6	Digital Museum for cultural and natural exhibits of the municipality (even if they are located somewhere else)	Thessaloniki, Trikala
8. Economy - Development		
8.1	Actions promoting entrepreneurship on municipal website	Thessaloniki, Amsterdam

#	Developments	Cities Examples
8.2	Actions for reinforcement, promotion, sale of local products via municipal website	Samos, Hamilton,
8.3	Employment actions via municipal website	Thessaloniki
8.4	Innovative actions to support high tech farming (eg. precision farming in municipal fields)	
8.5	Emergence - promotion of innovative technological activities via municipal website	Thessaloniki
8.6	Consulting interactive services to young entrepreneurs on municipal web platform	Thessaloniki
8.7	Crowdsourcing competitions platform	Thessaloniki
8.8	Contactless fare payment system	Singapore
8.9	Crypto Currencies adoption. // City wallet to foster local economy	
9. Security		
9.1	Early warning system and response to fires, floods, emergencies (+Fire detection systems based on drone surveillance)	Seville, Petropolis
9.2	Actions addressing citizens and protection plans on emergencies (i.e. earthquakes, floods)	Seville
9.3	Providence to guard public buildings – facilities	Seville
9.4	Monitoring weather conditions (weather forecast) for agricultural produce purposes	Amsterdam, Genoa
9.5	Events and Crowd management	Seville
10. E-Government		

#	Developments	Cities Examples
10.1	Electronic voting application (e-voting) for municipal issues	Heraklion
10.2	Application of Electronic Consultation on important decisions and plans (business plan, technical project, etc.)	Thessaloniki, Trikala, Lublin,
10.3	Collection of electronic signatures on important issues concerning citizens	Thessaloniki, Trikala
10.4	Converting the provision of municipal services to online in order to be accessible to all citizens via municipal website	Thessaloniki
10.5	Developing applications for citizens for their problems and requests reporting	Heraklion, Trikala
10.6	Online monitoring system of collective bodies of municipal meetings	Heraklion
10.7	Free access to open data for use by other public agencies or individuals	Barcelona, Thessaloniki
10.8	GIS applications for urban building construction, (suitable building sites, land use, objective property values etc.)	Barcelona, Thessaloniki, Heraklion
10.9	Implementation of eGovernment Services provision framework. On-line service delivery (eGov 1.0) including 311 systems	New York, Panama City
10.10	Systems to capture citizen Sentiment through social media	Delft
10.11	Metrics to evaluate city and Government performance	Lublin

Table reproduced with permission from the Authors:

Charalampos Alexopoulos	University of the Aegean, Mytilene, Greece
Gabriela Viale Pereira	Danube University Krems, Krems, Austria
Yannis Charalabidis	National Technical University of Athens, Athina, Greece
Lorenzo Madrid	Smart City Business Institute, FL, United States

List of Figures

List of Tables

References

[1] The Wealth of Cities, PGIM, 2015, p.1, http://www3.prudential.com/woc/_resources/media/the-wealth-of-cities.pdf, viewed 17 January 2016.

[2] City Mayors Statistics, "The largest cities in the world by land area, population and density, undated, http://www.citymayors.com/statistics/largest-cities-population-125.html, viewed 31 January 2016.

[3] Smart Cities Council, Our Vision, John DeKeles, April 10, 2012, http://smartcitiescouncil.com/article/our-vision, viewed 19 January 2016.

[4] Market Research Future, "Smart City Market Research Report—Global Forecast 2022," June 2017, https://www.marketresearchfuture.com/reports/smart-city-market-2624 viewed 31 May 2019.

[5] Smart Cities Market, "Smart Cities Market worth $717.2 billion by 2023, undated, https://www.marketsandmarkets.com/PressReleases/smart-cities.asp, viewed 31 May 2019.

[6] Global Industry Analysis, Inc., Need to Sustainably Accommodate the Growing Urban Population & Increased Application of Artificial Intelligence in Smart City Projects to Spur growth of Smart City Technologies, May 20199, https://www.strategyr.com/MarketResearch/ViewInfoGraphNew.asp?code=MCP-7080, viewed 2 June, 2019.

[7] Smart Cities Association, "Global smart cities market to reach a whopping $3.5 trillion by 2026, undated, https://www.smartcitiesassociation.org/index.php/media-corner/news/1-global-smart-cities-market-to-reach-a-whopping-3-5-trillion-by-2026, viewed 31 May 2019.

[8] Charalampos Alexopoulos of the University of the Aegean, Yannis Charalabidis of the National Technical University of Athens, Gabriela Viale Pereira of Danube University Krems, and Lorenzo Madrid of the Smart City Business Institute, "A Taxonomy of Smart City Initiatives," ICEGOV, Sydney, Australia, February 20-22, 2019. https://dl.acm.org/citation.cfm?id=3326402

[9] United Nations, *E-Government Survey 2016*, New York, 2016, p. 129.

[10] Mauro D. Rios, "En busca de una definición de Gobierno Electrónico," 23 October 2014, https://novagob.org/en-busca-de-una-definicia3n-de-gobierno-electra3nico/, viewed 8 April 2018.

[11] Gertrude Ntulo and Japhet Otike, "E-Government: Its Role, Importance and Challenges," undated, https://www.researchgate.net/file.PostFileLoader.html?id=564b965d622

5ffe6e98b4595&assetKey=AS:296884838125570@1447794269180, viewed 30 June 2018.

[12] Organisation for Economic Co-Operation and Development, *The e-Government Imperative*, Paris, 2003, https://read.oecd-ilibrary.org/governance/the-e-government-imperative_9789264101197-en#page1, viewed 27 April 2018.

[13] Organisation for Economic Co-Operation and Development, *The e-Government Imperative*, Paris, 2003, https://read.oecd-ilibrary.org/governance/the-e-government-imperative_9789264101197-en#page1, viewed 27 April 2018.

[14] https://dl.acm.org/citation.cfm?id=3326402 viewed 14 June 2019

[15] https://mind42.com/mindmap/f43c65b9-cb32-4bc7-b453-930bd5efc4be?rel=pmb

[16] The EA Pad, "White Paper on Enterprise Architecture 2003, Danish White Paper on Enterprise Architecture, https://eapad.dk/gov/dk/white-paper-on-enterprise-architecture/, viewed 26 January 2016.

[17] https://www.smartcities.at/assets/03-Begleitmassnahmen/Kurzfassung-SC-STANDARDS-engl.pdf

[18] https://www.bsigroup.com/en-IN/smart-cities/Smart-Cities-Standards-and-Publication/

[19] AENOR, *Spanish Standardization on Smart Cities*, 2015, https://portal.aenormas.aenor.com/descargasweb/normas/aenor-Spanish-standardization-on-Smart-Cities-CTN-178.pdf, viewed 17 May 2019.

[20] NIST, Smart City Framework Library, https://pages.nist.gov/smartcitiesarchitecture/library/, viewed 17 May 2019.

[21] Theresa A. Pardo and Brian G. Burke, *Improving Government Interoperability: A capability framework for government managers*, Center for Technology in Government, University at Albany, SUNY, 2008.

[22] Michael Munger, Library Economics Liberty, "Planning Order, Causing Chaos: Transantiago, "2008.

[23] Global Mass Transit Report, "Transantiago reforms, Chile: Preparing for Version 2.0, December 1, 2017, https://www.globalmasstransit.net/archive.php?id=28637, viewed 13 May 2019.

[24] The White House, Office of Management and Budget, Office of E-Government & Information, undated, https://obamawhitehouse.archives.gov/omb/e-gov, viewed 23 April 2018.

[25] U.S. Department of Health and Human Services, Office of Inspector General, *Healthare.gov*, Washington, D.C., February 2016, p.ii..

[26] U.S. Department of Health and Human Services, Office of Inspector General, *Healthare.gov*, Washington, D.C., February 2016, https://oig.hhs.gov/oei/reports/oei-06-14-00350.pdf, viewed 24 April 2018.

[27] *Hong Kong Smart City Blueprint*, Innovation and Technology Bureau, December 2017, https://www.smartcity.gov.hk/doc/HongKongSmartCityBlueprint(EN).pdf, viewed 17 May 2019.

[28] EasyPark, 2017 Smart Cities Index, https://www.easyparkgroup.com/smart-cities-index/, viewed 17 May 2019.

[29] Yupina Ng, "What's holding Hong Kong back from becoming a smarter city?" South China Morning Post, 24 February 2018, https://www.scmp.com/news/hong-kong/community/article/2134354/whats-holding-hong-kong-back-becoming-smarter-city, viewed 17 May 2019.

[30] Yam Ki Chan, "Hong Kong's smart city ambitions must be powered by tech-savvy people," *South China Morning Post*, 3 October, 2018, https://www.scmp.com/comment/insight-opinion/hong-kong/article/2166773/hong-kongs-smart-city-ambitions-must-be-powered, viewed 1 May 2019.

[31] OECD E-Government Project, *Draft OECD Principles on Digital Government Strategies: Bringing Governments Closer to Businesses*, 2013, http://www.oecd.org/governance/eleaders/Draft-OECD-Principles-for-Digital-Government-Strategies.pdf, viewed 6 June 2019.

[32] Jack J. Phillips, Ph.D., "ROI in the Public Sector: Myths and Realities," originally published in Public Personnel Management, June 22, 2004, https://roiinstitute.net/wp-content/uploads/2017/02/ROI-in-Public-Sector-Myths-and-Realities.pdf, viewed 12 July 2019.

[33] Tallinn Ministerial Declaration on eGovernment, 6 October 2017, http://ec.europa.eu/newsroom/document.cfm?doc_id=47559, viewed 20 April 2018.

[34] Gianluigi Viscusi, Carlo Batini, and Massimo Mecella, *Information Systems for eGovernment*, Springer, Heidelberg, 2010, p.

[35] United Nations Development Programme, *e-Government Interoperability: Guide*, Bangkok, 2007, pp.13-19.

[36] United Nations, *E-Government Survey 2016*, New York, 2016, p.57.

[37] Republic of Bulgaria, e-Governance Development Strategy 2014-2020 in the Republic of Bulgaria, undated, https://www.mtitc.government.bg/sites/default/files/uploads/pdf/e_governance_strategy.pdf, viewed 3 May 2018.

[38] Republic of Bulgaria, e-Governance Development Strategy 2014-2020 in the Republic of Bulgaria, undated, https://www.mtitc.government.bg/sites/default/files/uploads/pdf/e_governance_strategy.pdf, viewed 3 May 2018.

[39] European Commission, eGovernment in Bulgaria, 2016, https://joinup.ec.europa.eu/sites/default/files/inline-files/eGovernment%20in%20Bulgaria%20-%20February%202016%20-%2013_0%20-%20v3_00.pdf, viewed 3 May 2018.

[40] City of Boston, Digital Strategy, http://www.cityofboston.gov/DoIT/strategy/, viewed 11 April 2016.

[41] City of Boston, About the Department of Innovation and Technology, http://www.cityofboston.gov/DoIT/about/, viewed 11 April 2016.

[42] https://www.wmur.com/article/sensors-placed-on-memorial-bridge-by-unh-engineers-collecting-structure-environment-data/27735886 as viewed on June 6, 2019

[43] J. Barton Cunningham and James S. Kempling, "Implementing change in public sector organizations," *Management Decision*, Vol. 47, No. 2, 2009, pp. 330-244, https://pdfs.semanticscholar.org/e1a0/0713d92040a80790c0112747545bc5685e9e.pdf, viewed 13 July 2019.

[44] Workforce, "Successful Change Management Practices in the Public Sector," undated, https://www.govexec.com/assets/sponsored-change-management-practices.pdf?registration-funnel=ge-mod-conv-dt, viewed 13 July 2019.

[45] https://www.service-architecture.com/articles/web-services/service-oriented_architecture_soa_definition.html

[46] https://en.wikipedia.org/wiki/General_Data_Protection_Regulation

[47] Microoft Azure, "What is middleware?" undated, https://azure.microsoft.com/en-us/overview/what-is-middleware/, viewed 7 June 2019,

[48] https://s3.amazonaws.com/nist-sgcps/cpspwg/files/pwgglobal/CPS_PWG_Framework_for_Cyber_Physical_Systems_Release_1_0Final.pdf

[49] NIST SP 1500, Vol. 6 "Big Data Interoperability Framework: Reference Architecture"

[50] https://share.ansi.org/Shared%20Documents/News%20and%20Publications/IES-CityFramework_Version_1_0_20180930.pdf

[51] John Pearson, "Thoughts about Decision Making in Government, American Society for Public Administration, 12 February 2018, https://patimes.org/thoughts-decision-making-government/, viewed 21 May 2018.

[52] Leif Sundberg and Aron Larsson, "The Impact of Formal Decision Processes on e-Government Projects," Administrative Sciences, 22 May 2017.

[53] Mahesh Kelkar, et al., "Mission analytics," Deloitte, 26 September 2016, https://www2.deloitte.com/insights/us/en/industry/public-sector/data-driven-decision-making-in-government.html, viewed 21 May 2018.

[54] FinTech, Blockchain and the notaries: the services won't be replaced by transformed, 15 August 2018, https://www.bankingtech.com/2018/08/blockchain-and-the-notaries-the-services-wont-be-replaced-but-transformed/, viewed 23 June 2019.

[55] KC Tam, Notarization in Blockchain (Part 1), August 28, 2018, https://medium.com/@kctheservant/notarization-in-blockchain-part-1-a9795f19e28d, viewed 23 June 2019.

[56] Lexology, Using blockchain for land registry, undated, https://www.lexology.com/library/detail.aspx?g=36641e13-b81d-48fc-95e9-3ea3bc9135aa, viewed 23 June 2019.

[57] David Hamilton, Blockchain Land Registry: The New Kid on the Block,, Coin Central, 11 January 2019, In fact, by enabling the parties to have the necessary data in a timely manner, all sorts of contracts could be streamlined to be faster and more efficient, viewed 23 June 2019.

[58] Blaž Golob, Smart City Platform: Empowering People with Digital Services, GoForeSight Institute & SmartIS, undated, Ig & Ljubljana, Slovenia.

[59] Cisco, "Smart Cities are Turning Data into Action," undated, http://www.cisco.com/c/m/en_us/never-better/core-networking2.html?POSITION=SEM&COUNTRY_SITE=us&CAMPAIGN=NBT-07+Networking&CREATIVE=SEM_NB_Smart_Cities_NBT_(BMM)_(NB)-Public_Network_LPTEST2&REFERRING_SITE=Google&KEYWORD=%2Bsmart%20%2Bcities&KWID=p12556153967&gclid=CNaRiOyky84CFZWIaQod5eYMoQ&dclid=CKGzreyky84CFVNCAQodOGgDyQ, viewed 18 August 2016.

[60] Christopher Frey, "World Cup 2014: Inside Rio's Bond-villain mission control," The Guardian, 23 May 2014, https://www.theguardian.com/cities/2014/may/23/world-cup-inside-rio-bond-villain-mission-control, viewed 19 August 2016.

[61] ASTRO Systems, Technical Assistance for Istanbul Smarter City Initiative.

[62] Smart Cities Council, *Smart Street Lighting 101*, undated, Forward.

[63] U.S. Department of Energy, Office of Electricity Delivery & Energy Reliability, Smart Grid, undated, http://energy.gov/oe/services/technology-development/smart-grid, viewed 19 May 2016.

[64] Smart Cities Council, *Smart Street Lighting 101*, undated, p.5.

[65] Cisco, Smart+Connected Parking, undated, http://www.cisco.com/c/en/us/solutions/industries/smart-connected-communities/city-parking.html, viewed 18 August 2016.

[66] Smart Cities Council, Smart Cities Readiness Guide,® Public Safety, undated, http://readinessguide.smartcitiescouncil.com/readiness-guide/public-safety-0, viewed 3 September 2016.

[67] Lorenzo Madrid and Linda Lee Bower, "How to Build and Manage Smart Cities: an Open Framework Proposal," Samos 2016 Summit on ICT-Enabled Governance, Samos, Greece, July 4, 2016.

[68] Bel Williamson, "Educating the smart city: Schooling smart citizens through computational urbanism, undated, http://bds.sagepub.com/content/spbds/2/2/2053951715617783.full.pdf, viewed 28 August 2016.

[69] V. Mayer-Schönberger and K. Cukier, *Learning from Big Data: the Future of Education*, Houghton Mifflin Harcourt Publishing Co., New York, 2014.

[70] MK:Smart, http://www.mksmart.org/, viewed 1 June 2019.

[71] Urban Data School Aims, 2014, http://urbandataschool.org.

[72] Medic-Mobile, http://medicmobile.org/, viewed 19 August, 2016.

[73] United Nations, *E-Government Survey 2016*, New York, 2016, p.iii.

[74] United Nations, *E-Government Survey 2016*, New York, 2016, , p.11.

[75] Cresswell, Anthony M., et al., *Developing Public Value Metrics for Returns to Government ICT Investments*, Center for Technology in Government, University at Albany-SUNY, June 2012.

[76] United States General Accounting Office, Accounting and Information Management Division, *Assessing Risks and Returns: A guide for Evaluating Federal Agencies' IT Investment Decision-making*, Washington, D.C., Version 1, February 1997, p.1.

[77] ASTRO Systems, Inc., Feasibility Study, Connectivity Infrastructure Project for the Government of Costa Rica, Deliverable No. 5, *Economic and Financial Analysis* (Task 7)

[78] Lorenzo Madrid, *The Economic Impact of Interoperability, Microsoft, 2005.*

[79] Lorenzo Madrid, Director, Technology Office Strategy, Microsoft, WW Public Sector, *The Challenge of Interoperability in delivering Citizens Services*, March 2011.

[80] ISO, Standards catalogue, ISO 9001:2015, Quality management systems—Requirements, Abstract, undated, http://www.iso.org/iso/home/store/catalogue_tc/catalogue_detail.htm?csnumber=62085, viewed 15 January 2016/

[81] ISO 18091:2014, https://www.iso.org/obp/ui/#iso:std:iso:18091:e. d-1:v1:en, viewed 15 January 2016.

[82] ISO 37120:2014(en), https://www.iso.org/obp/ui/#iso:std:iso:37120:ed-1:v1:en, viewed 15 January 2016.

[83] International Standards Organization, ISO 37122:2019, published July 2019, https://www.iso.org/standard/69050.html, viewed 13 July 2019.

[84] Social Progress Imperative, Social Progress Index, http://www.socialprogressimperative.org/data/spi, viewed 16 January 2016.
[85] Social Progress Imperative, About, http://www.socialprogressimperative.org/about/the-imperative, viewed 17 January 2016.
[86] #Progreso Social Colombia, *Social Progress Index for Cities of Colombia*, Social Progress Imperative, Executive Summary, 2016, https://www2.deloitte.com/content/dam/Deloitte/global/Documents/About-Deloitte/gx-dttl-spi-for-cities-in-colombia.pdf, viewed 14 May 2019.
[87] GreenTech Media, Energy Efficiency, "Schneider Electric Brings Smart City Tech to Boston," Jeff St. John, 18 September 2013, http://www.greentechmedia.com/articles/read/schneider-electric-brings-smart-city-tech-to-boston, viewed 3 January 2016.
[88] *United Nations E-Government Survey 2018*,
[89] Nokia, *Smart City Playbook.*
[90] United Nations, *Smart City Strategy in South Africa*, Case Study Overview, undated, http://unpan1.un.org/intradoc/groups/public/documents/cpsi/unpan033820.pdf, viewed 28 August 2018
[91] City of Cape Town, *Five-Year Plan for Cape Town 2012-2017*, https://www.westerncape.gov.za/text/2012/11/city-of-cape-town-idp-2012-2017.pdf, viewed 10 May 2018.
[92] City of Cape Town, *Five-Year Integrated Development Plan July 2017-June 2022*, http://resource.capetown.gov.za/documentcentre/Documents/City%20strategies%2c%20plans%20and%20frameworks/IDP%202017-2022.pdf, viewed 10 May 2019
[93] https://news.sap.com/2017/01/sap-cape-town-emergency-management-epic/ viewed 1 July 2019
[94] Gutierrez, Peter. "Using LoRa to build a 'Fitbit' for Sydney," IoTHub, 20 July 2016, https://www.iothub.com.au/news/using-lora-to-build-afitbit- for-sydney-431102, viewed 1 September 2018
[95] City of Sydney, Digital Sydney, https://www.cityofsydney.nsw.gov.au/__data/assets/pdf_file/0005/288167/Digital-Strategy.pdf, viewed 11 May 2019.
[96] Commonwealth of Australia, Department of the Prime Minister and Cabinet, *Smart Cities Plan*, 2016.
[97] Republic of Turkey, Ministry of Development, *The Tenth Development Plan 2014-2018*, Ankara, 2014, http://www.mod.gov.tr/Lists/RecentPublications/Attachments/75/The%20Tenth%20Development%20Plan%20(2014-2018).pdf, viewed 1 September 2016.

[98] AA, "Turkish vice president introduces 11th development plan," 9 July 2019, https://www.aa.com.tr/en/economy/turkish-vice-president-introduces-11th-development-plan/1527269, viewed 18 July 2019.

[99] Istanbul Development Agency, 2014-2023 Istanbul Regional Plan, http://www.istka.org.tr/content/pdf/Istanbul-Regional-Plan-2014-2023.pdf, viewed 1 September 2016.

[100] United States Trade and Development Agency, Technical Assistance for Istanbul Smarter City Initiative, carried out by ASTRO Systems, 2016.

[101] City SDK, https://www.citysdk.eu/.

[102] Shanghai, Action Plan 2011-2013 of Shanghai Municipality for Building Smart City, 2013, http://www.shanghai.gov.cn/shanghai/node27118/node27973/u22ai708 98.html, viewed 27 May 2019.

[103] Xueguo Wen, Charting the Progress of Smart City Development in Shanghai, presentation, 2017, https://dtw.tmforum.org/wp-content/uploads/2017/05/3.-Professor-Wen-Xueguo.pdf, viewed 25 May 2-2019.

[104] Chia Jie Lin, "Five Chinese Smart cities leading the way," Government Insider, 10 July 2018, https://govinsider.asia/security/five-chinese-smart-cities-leading-way/, viewed 26 May 2019.

[105] Ayuntamiento de Sevilla, *Introducción*, undated, https://www.sevilla.org/ayuntamiento/competencias-areas/alcaldia/itas/plan-director-de-innovacion/doc/introduccion.pdf, viewed 14 May 2018.

[106] FIWARE, FIWARE: The Open Source Platform for our Smart Digital Future, undated, https://www.fiware.org/, viewed 16 May 2018.

[107] Opensource.com, "Open source FIWARE platform creates new IoT business opportunities," 9 November 2016, https://opensource.com/business/16/11/fiware-platform, viewed 16 May 2018,.

[108] Wellness Telecom, Successful Cases, undated, http://www.wtelecom.es/casos/caso-de-estudio-life-ewas-wellness-smart-cities-lipasam/, viewed 17 May 2018.

Made in the USA
Middletown, DE
07 February 2022

60710876R00088